厦门理工学院教材建设基金资助

Professional English for Electrical Engineering & Automation

电气工程与自动化专业英语

主　编：郑雪钦

副主编：苏鹭梅　高海燕

厦门大学出版社 XIAMEN UNIVERSITY PRESS

国家一级出版社

全国百佳图书出版单位

图书在版编目(CIP)数据

电气工程与自动化专业英语/郑雪钦主编.—厦门:厦门大学出版社,2018.10
ISBN 978-7-5615-7076-0

Ⅰ.①电… Ⅱ.①郑… Ⅲ.①电气工程—英语—高等学校—教材②自动化技术—英语—高等学校—教材 Ⅳ.①TM②TP2

中国版本图书馆 CIP 数据核字(2018)第 207338 号

出 版 人 郑文礼
责任编辑 郑 丹
封面设计 蒋卓群
技术编辑 许克华

出版发行 厦门大学出版社
社 址 厦门市软件园二期望海路 39 号
邮政编码 361008
总 编 办 0592-2182177 0592-2181406(传真)
营销中心 0592-2184458 0592-2181365
网 址 http://www.xmupress.com
邮 箱 xmupress@126.com
印 刷 厦门市金凯龙印刷有限公司

开本 787 mm×1 092 mm 1/16
印张 10.5
字数 300 千字
版次 2018 年 10 月第 1 版
印次 2018 年 10 月第 1 次印刷
定价 32.00 元

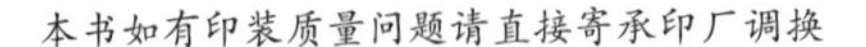
本书如有印装质量问题请直接寄承印厂调换

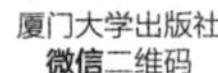
厦门大学出版社
微信二维码

厦门大学出版社
微博二维码

前　言

通过专业英语的学习，可以培养学生对英文专业资料、文献和信息的阅读能力、专业写作能力，扩充学生的专业词汇量，帮助学生掌握科技英语和专业英语的特点。

本书供普通高等院校电气类和自动化本科专业使用，共六章。分别为第一章“电子电路”，第二章“电机学”，第三章“电力电子”，第四章“电气工程”，第五章“自动化”，第六章“智能电网信息技术”。每章结合专业英语特点，增设课堂实践，如文献检索、视听口译、器件说明书阅读、科研报告的幻灯片制作及其演说、论文摘要的撰写等练习，力求理论与实践的深度融合。

本书第一章和第三章由苏鹭梅编写，第二章和第四章由郑雪钦编写，第五章和第六章由高海燕编写。全书由郑雪钦和苏鹭梅统稿。本书大部分插图绘制、文稿录入、资料收集等工作由研究生陈晓雄、吴景丽和陈明泉完成，在此表示衷心的感谢。

本书的出版得到厦门理工学院教材建设基金项目资助，在此表示衷心的感谢。

在本书完稿之际，对书末所附参考文献的作者也致以衷心的感谢。

笔者虽然长期工作在专业英语的教学第一线，但毕竟水平有限，加之编写时间仓促，书中难免有些疏漏及错误，请读者批评指正。

编　者

于厦门理工学院电气工程与自动化学院

2018 年 5 月

目　录

CONTENTS

Chapter 1

ELECTRIC CIRCUIT

Unit 1 Basic Circuit Elements

1.1 Electrical Resistance (Ohm's Law)

Resistance is the capacity of materials to impede the flow of current, or more specifically, the flow of electric charge. The circuit element used to model this behavior is the resistor. Fig. 1-1 shows the circuit symbol for the resistor, with R denoting the resistance value of the resistor.

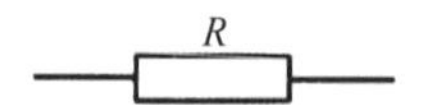

Fig. 1-1 The circuit symbol for a resistor having a resistance R

Conceptually, we can understand resistance if we think about the moving electrons that make up electric current interacting with and being resisted by the atomic structure of the material through which they are moving. In the course of these interactions, some amount of electric energy is converted to thermal energy and dissipated in the form of heat. This effect may be undesirable. However, many useful electrical devices take advantage of resistance heating, including stoves, toasters, irons, and space heaters.

1.2 The Capacitor

1.2.1 Ideal Capacitor Model

Previously, we termed independent and dependent sources active elements, and the linear resistor a passive element, although our definitions of active and passive are still slightly fuzzy and need to be brought into sharper focus. We now define an active element as an element that is capable of furnishing an average power greater than zero to some external device, where the average is taken over an infinite time interval. Ideal sources are active elements, and the operational amplifier is also an active device. A passive element however, is defined as an element that cannot supply an average power that is greater than zero over an infinite time interval. The resistor falls into this category; the energy it receives is usually transformed into heat, and it never supplies energy.

We now introduce a new passive circuit element, the capacitor. We define capacitance C by the voltage-current relationship as in Eq. (1-1), where v and i satisfy the conventions for a passive element, as shown in Fig. 1-2.

$$i=C\frac{\mathrm{d}v}{\mathrm{d}t} \tag{1-1}$$

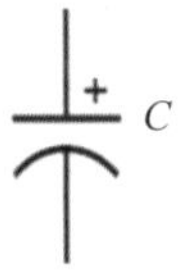

Fig. 1-2 Electrical symbol and current-voltage conventions for a capacitor

We should bear in mind that v and i are functions of time. If needed, we can emphasize this fact by writing $v(t)$ and $i(t)$ instead. From Eq. (1-1), we may determine the unit of capacitance as an ampere-second per volt, or coulomb per volt. We will now define the farad (F) as one coulomb per volt, and use this as the unit of capacitance.

The ideal capacitor defined by Eq. (1-1) is only a mathematical model of a real device. A capacitor consists of two conducting surfaces on which charge may be stored, separated by a thin insulating layer that has a very large resistance. If we assume that this resistance is sufficiently large that may be considered infinite, then equal and opposite charges placed on the capacitor "plates" can never recombine, at least by any path within the element. The construction of the physical device is suggested by the circuit symbol shown in Fig. 1-2.

1.2.2 Important Characteristics of an Ideal Capacitor

(1) There is no current through a capacitor if the voltage across it is not changing with time. A capacitor is therefore an open circuit to DC.

(2) A finite amount of energy can be stored in a capacitor even if the current through the capacitor is zero, such as when the voltage across it is constant.

(3) It is impossible to change the voltage across a capacitor by a finite amount in zero time, as this requires an infinite current through the capacitor. (A capacitor resists an abrupt change in the voltage across it in a manner analogous to the way a spring resists an abrupt

change in its displacement.)

(4) A capacitor never dissipates energy, but only stores it. Although this is true for the mathematical model, it is not true for a physical capacitor due to finite resistances associated with the dielectric as well as the packaging.

1.3 The Inductor

In the early 1800s the Danish scientist Oersted showed that a current-carrying conductor produced a magnetic field (compass needles were affected in the presence of a wire when current was flowing). Shortly thereafter, Ampere made some careful measurements which demonstrated that this magnetic field was linearly related to the current which produced it. The next step occurred some 20 years later when the English experimentalist Michael Faraday and the American inventor Joseph Henry discovered almost simultaneously that a changing magnetic field could induce a voltage neighboring circuit. They showed that this voltage was proportional time rate of change of the current producing the magnetic field. The constant of proportionality is what we now call the inductance, symbolized by L and therefore:

$$v = L\frac{\mathrm{d}i}{\mathrm{d}t} \tag{1-2}$$

We must realize that v and i are both functions of time. When we wish to emphasize this, we may do so by using the symbols $v(t)$ and $i(t)$.

The circuit symbol for the inductor is shown in Fig. 1-3, and it should be noted that the passive sign convention is used, just as it was with the resistor and the capacitor. The unit in which inductance is measured is the henry (H), and the defining equation shows that the henry is just a shorter expression for a volt-second per ampere.

Fig. 1-3 Electrical symbol and current-voltage conventions for an inductor

1.4 Voltage and Current Sources

Before discussing ideal voltage and current sources, we need to consider the general nature of electrical sources. An electrical source is a device that is capable of converting nonelectric energy to electric energy and vice versa. A discharging battery converts chemical energy to electric energy, whereas a battery being charged converts electric energy to chemical energy. A dynamo is a machine that converts mechanical energy to electric energy and vice versa. If operating in the mechanical-to-electric mode, it is called a generator. If transforming from electric to mechanical energy, it is referred to as a motor. The important thing to remember about these sources is that they can either deliver or absorb electric power, generally maintaining either voltage or current. This behavior is of particular interest for circuit analysis and led to the creation of the ideal voltage source and the ideal current source as basic circuit elements. The challenge is to model practical sources in terms of the ideal basic circuit elements.

An ideal voltage source is a circuit element that maintains a prescribed voltage across its terminals regardless of the current flowing in those terminals. Similarly, an ideal current source is a circuit element that maintains a prescribed current through its terminals regardless of the voltage across those terminals. These circuit elements do not exist as practical devices — they are idealized models of actual voltage and current sources.

Using an ideal model for current and voltage sources places an important restriction on how we may describe them mathematically. Because an ideal voltage source provides a steady voltage, even if the current in the element changes, it is impossible to specify the current in an ideal voltage source as a function of its voltage. Likewise, if the only information you have about an ideal current source is the value of current supplied, it is impossible to determine the voltage across that current source. We have sacrificed our ability to relate voltage and current in a practical source for the simplicity of using ideal sources in circuit analysis.

Specialized English Words

circuit element 电路元件
electric charge 电荷
electron 电子
capacitor 电容器
operational amplifier 运算放大器
dielectric 电介质
conductor 导体
wire 导线
generator 发电机
resistance 电阻
resistor 电阻器
electric energy 电能
active/passive element 有源 / 无源元件
insulating layer 绝缘层
inductor 电感器
magnetic field 磁场
inductance 电感
motor 马达；电动机

Unit 2 Basic Circuit Laws

2.1 Introduction

We are now ready to meet another idealized element, the linear resistor. The resistor is the simplest passive element, and we begin our discussion by considering the work of an obscure German physicist, Georg Simon Ohm, who published a pamphlet in 1827 that described the results of one of the first efforts to measure currents and voltages, and to describe and relate them mathematically. One result was a statement of the fundamental relationship we now call Ohm's law, even though it has since been shown that this result was discovered 46 years earlier in England by Hey Cavendish, a brilliant semi-recluse.

Ohm's law states that the voltage across conducting materials is directly proportional to the current flowing through the material as defined in Eq. (1-3), where the constant of proportionality R is called the resistance. The unit of resistance is the ohm, which is 1 V/A and customarily abbreviated by a capital omega, Ω.

$$v = Ri \tag{1-3}$$

2.2 Kirchhoff's Current Law

We are now ready to consider the first of the two laws named for Gustav Robert Kirchhoff (two h's and two f's), a German university professor who was born about the time Ohm was doing his experimental work. This axiomatic law is called Kirchhoff's current law (abbreviated KCL), and it simply states that

The algebraic sum of the currents entering any node is zero.

This law represents a mathematical statement of the fact that charge cannot accumulate at a node. A node is not a circuit element, and it certainly cannot store, destroy, or generate charge. Hence, the currents must sum to zero. A hydraulic analogy is sometimes useful here: for example, consider three water pipes joined in the shape of a Y. We define three "currents" flowing into each of the three pipes. If we insist that water is always flowing, then obviously we cannot have three positive water currents, or the pipes would burst. This is a result of our defining currents independent of the direction that water is actually flowing. Therefore, the value of either one or two of the currents as defined must be negative. Consider the node shown in Fig. 1-4. The algebraic sum of the four currents entering the node must be zero, as defined by Eq. (1-4).

$$i_A + i_B + (-i_C) + (-i_D) = 0 \tag{1-4}$$

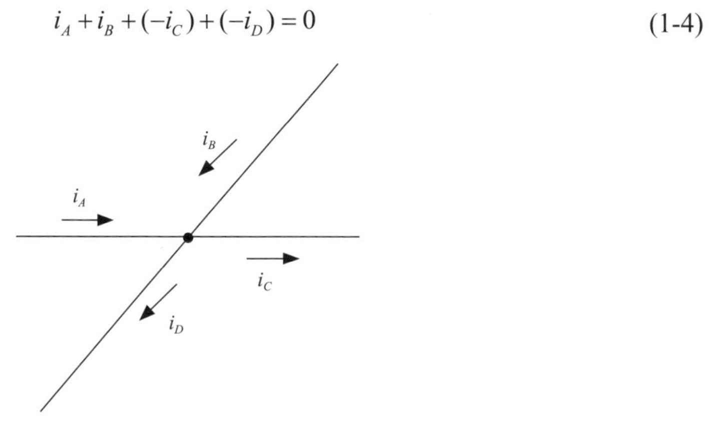

Fig. 1-4 Example node to illustrate the application of Kirchhoff's current law

However, the law could be equally well applied to the algebraic sum of the currents leaving the node, as defined by Eq. (1-5).

$$(-i_A) + (-i_B) + i_C + i_D = 0 \tag{1-5}$$

We might also wish to equate the sum of the currents having reference arrows directed into the node to the sum of those directed out of the node, which simply states that the sum of the currents going in must equal the sum of the currents going out, as defined by Eq. (1-6).

$$i_A + i_B = i_C + i_D \tag{1-6}$$

2.3 Kirchhoff's Voltage Law

Current is related to the charge flowing through a circuit element, whereas voltage is a measure of potential energy difference across the element. There is a single unique value for any voltage in circuit theory. Thus, the energy required to move a unit charge from point *A* to point *B* in a circuit must have a value independent of the path chosen to get from *A* to *B* (there is often more than one such path). We may assert this fact through Kirchhoff's voltage law (abbreviated KVL):

The algebraic sum of the voltages around any closed path is zero.

In Fig. 1-5, if we carry a charge of 1 C from *A* to *B* through element the reference polarity signs for v_1 show that we do v_1 joules of work. Now if, instead, we choose to proceed from *A* to *B* via node *C*, then we expend $(v_2 - v_3)$ joules of energy. The work done, however, is independent of the path in a circuit, and so any route must lead to the same value for the voltage. In other words,

$$v_1 = v_2 - v_3 \tag{1-7}$$

It follows that if we trace out a closed path, the algebraic sum of the voltages across the individual elements around it must be zero. Thus, we may write

$$v_1+v_2+v_3+\cdots+v_N=0 \tag{1-8}$$

or, more compactly,

$$\sum_{n=1}^{N} v_n=0 \tag{1-9}$$

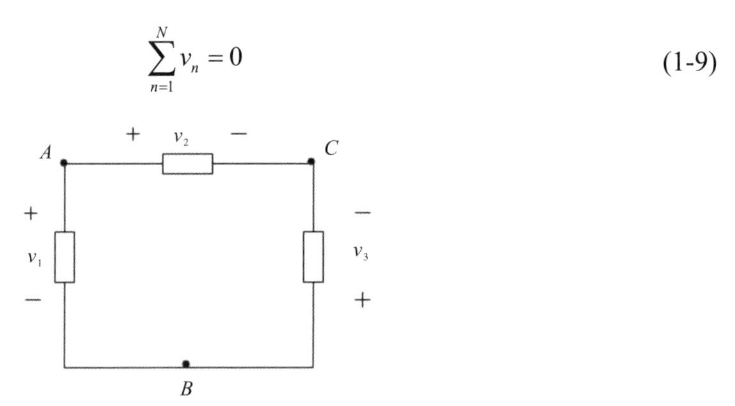

Fig. 1-5 Illustration for the potential difference between points *A* and *B* is independent of the path selected

Specialized English Words

passive element 无源元件

resistance 电阻

charge 电荷

polarity 极性

joule 焦耳

conducting material 导电材料，导电物质

Kirchhoff's Current Law (KCL) 基尔霍夫电流定律

hydraulic 液压的，水力的

Kirchhoff's Voltage Law (KVL) 基尔霍夫电压定律

Unit 3 Basic Circuit Analysis Methods

3.1 Introduction

Armed with the trio of Ohm's and Kirchhoff's laws, analyzing a simple linear circuit to obtain useful information such as the current, voltage, or power associated with a particular element is perhaps starting to seem a straightforward enough venture. Still, for the moment at least, every circuit seems unique, requiring (to some degree) a measure of creativity in approaching the analysis. In this Unit, we learn two basic circuit analysis techniques—**nodal analysis** and **mesh analysis**—both of which allow us to investigate many different circuits with a consistent, methodical approach. The result is a streamlined analysis, a more uniform level of complexity in our equations, fewer errors and, perhaps most importantly, a reduced occurrence of "I don't know how to even start!"

3.2 Nodal Analysis

We begin our study of general methods for methodical circuit analysis by considering a powerful method based on KCL, namely nodal analysis. In other lessons we considered the analysis of a simple circuit containing only two nodes. We found that the major step of the analysis was obtaining a single equation in terms of a single unknown quantity—the voltage between the pair of nodes.

We will now let the number of nodes increase and correspondingly provide one additional unknown quantity and one additional equation for each added node. Thus, a three-node circuit should have two unknown voltages and two equations; a 10-node circuit will have nine unknown voltages and nine equations; an N-node circuit will need (N-l) voltages and (N-l) equations. Each equation is a simple KCL equation.

The mesh is a property of a planar circuit. We define a mesh as a loop that does not contain any other loop within it. Once a circuit has been drawn neatly in planar form, it often has the appearance of a multipaned window and the boundary of each pane in the window may be considered to be a mesh.

If a network is planar, mesh analysis can be used to accomplish the analysis. This technique involves the concept of a mesh current, which is introduced by considering the analysis of the two-mesh circuit.

As we did in the single-loop circuit, we will begin by defining a current through one of the branches. Let us call the current flowing to the right through the 6 Ω resistor i_1. We will apply KVL around each of the meshes, and the two resulting equations are sufficient to determine two unknown currents. We next define a second current i_2 flowing to the right of the 4 Ω resistor. We might also choose to call the current flowing downward through the central branch i_3, but it is evident from KCL that i_3 may be expressed in terms of the two previously assumed currents as ($i_1 - i_2$).

Following the method of solution for the single loop circuit, we now apply KVL to the left-hand mesh and obtain

$$-42 + 6i_1 + 3(i_1 - i_2) = 0$$

or

$$9i_1 - 3i_2 = 42 \tag{1-10}$$

Applying KVL to the right-hand mesh, we obtain

$$-3(i_1 - i_2) + 4i_2 - 10 - 0$$

or

$$-3i_1 + 7i_2 = 10 \tag{1-11}$$

Eq. (1-10) and (1-11) are independent equations; one cannot be derived from the other. With two equations and two unknowns, the solution is easily obtained:

$$i_1 = 6A \quad i_2 = 4A$$

and

$$(i_1 - i_2) = 2A \tag{1-12}$$

3.3 Summary of Basic Nodal Analysis Procedure

(1) **Count the number of nodes** (N).

(2) **Designate a reference node.** The number of terms in your nodal equations can be minimized by selecting the node with the greatest number of branches connected to it.

(3) **Label the nodal voltages** (there are $N-1$ of them).

(4) **Write a KCL equation for each of the non-reference nodes.** Sum the currents flowing into a node from sources on one side of the equation. On the other side, sum the currents flowing out of the node through resistors. Pay close attention to "−" signs.

(5) **Express any additional unknowns such as currents or voltages other than nodal voltages in terms of appropriate nodal voltages.** This situation can occur if voltage sources or dependent sources appear in our circuit.

(6) **Organize the equations**. Group terms according to nodal voltages.

(7) **Solve the system of equations for the nodal voltages** (there will be $N-1$ of them).

3.4 Summary of Basic Mesh Analysis Procedure

(1) **Determine if the circuit is a planar circuit**. If not, perform nodal analysis instead.

(2) **Count the number of meshes (**M**)**. Redraw the circuit if necessary.

(3) **Label each of the M mesh currents**. Generally, defining all mesh currents to flow clockwise results in a simpler analysis.

(4) **Write a KVL equation around each mesh**. Begin with a convenient node and proceed in the direction of the mesh current. Pay close attention to "−" signs. If a current source lies on the periphery of a mesh, no KVL equation is needed and the mesh current is determined by inspection.

(5) **Express any additional unknowns such as voltages or currents other than mesh**

currents in terms of appropriate mesh currents. This situation can occur if current sources or dependent sources appear in our circuit.

(6) **Organize the equations.** Group terms according to mesh currents.

(7) **Solve the system of equations for the mesh currents** (there will be M of them).

Specialized English Words

linear circuit 线性电路
mesh analysis 网孔分析法
branch 分支
loop 回路
nodal analysis 节点分析法
node 节点
resistor 电阻器

Unit 4 Operational Amplifier

4.1 Ideal and Practical Models

The concept of the operational amplifier (usually referred to as an op-amp) originated at the beginning of the Second World War with the use of vacuum tubes in DC amplifier designs developed by the George A. Philbrick Co. (Some of the early history of operational amplifiers is found in Williams, 1991). The op-amp was the basic building block for early electronic servomechanisms, for synthesizers, and in particular for analog computers used to solve differential equations. With the advent of the first monolithic integrated circuit (IC) op-amp in 1965 (the μA709, designed by the late Bob Widlar, then with Fairchild Semiconductor), the availability of op-amps was no longer a factor, while within a few years the cost of these devices (which had been as high as $200 each) rapidly plummeted to close to that of individual discrete transistors.

Although the digital computer has now largely supplanted the analog computer in mathematically intensive applications, the use of inexpensive operational amplifiers in instrumentation applications, in pulse shaping, in filtering, and in signal processing applications in general has continued to grow. There are currently many commercial manufacturers whose main products are high quality op amps. This competitiveness has ensured a marketplace featuring a wide range of relatively inexpensive devices suitable for use by electronic engineers, physicists, chemists, biologists, and almost any discipline that requires obtaining quantitative analog data from instrumented experiments.

Most operational amplifier circuits can be analyzed, at least for first-order calculations, by considering the op-amp to be an "ideal" device. For more quantitative information, however, and particularly when frequency response and DC offsets are important, one must refer to a more "practical" model that includes the internal limitations of the device. If the op amp is characterized by a really complete model, the resulting circuit may be quite complex, leading to rather laborious calculations. Fortunately, however, computer analysis using the program SPICE significantly reduces the problem to one of a simple input specification to the computer. Today, nearly all the op-amp manufacturers provide SPICE models for their line of devices, with excellent correlation obtained between the computer simulation and the actual measured results.

4.2 The Ideal Op-Amp

An ideal operational amplifier is a DC coupled amplifier having two inputs and normally one output (although in a few infrequent cases there may be a differential output). The inputs are designated as non-inverting (designated + or NI) and inverting (designated−or Inv.). The amplified signal is the differential signal v_ε between the two inputs, so that the output voltage as indicated in Fig. 1-6 is

$$v_{out}=A_{ol}(v_B-v_A) \tag{1-13}$$

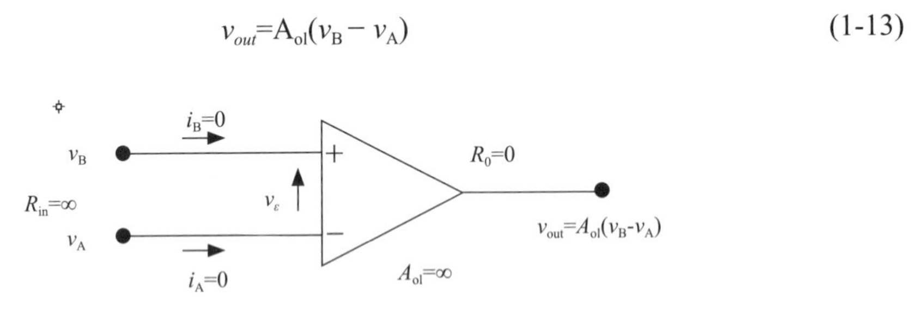

Fig. 1-6 Configuration for an ideal op amp

The general characteristics of an ideal op-amp can be summarized as follows:

(1) The open-loop gain A_{ol} is infinite. Or, since the output signal v_{out} is finite, then the differential input signal v_ε must approach zero.

(2) The input resistance R_{in} is infinite, while the output resistance R_0 is zero.

(3) The amplifier has zero current at the input (i_A and i_B in Fig. 1-6 are zero), but the op-amp can either sink or source an infinite current at the output.

(4) The op amp is not sensitive to a common signal on both inputs (i.e., $v_A = v_B$); thus, the output voltage change due to a common input signal will be zero. This common signal is referred to as a common mode signal, and manufacturers specify this effect by an op-amp as common mode rejection ratio (CMRR), which relates the ratio of the open-loop gain (A_{ol}) of

the op-amp to the common-mode gain (A_{cm}). Hence, for an ideal op amp CMRR = ∞.

(5) A somewhat analogous specification to the CMRR is the power supply rejection ratio (PSRR), which relates the ratio of a power supply voltage change to an equivalent input voltage change produced by the change in the power supply. Because an ideal op amp can operate with any power supply, without restriction, then for the ideal device PSRR = ∞.

(6) The gain of the op-amp is not a function of frequency. This implies an infinite bandwidth.

Although the foregoing requirements for an ideal op-amp appear to be impossible to achieve practically, modern devices can quite closely approximate many of these conditions. An op-amp with a field effect transistor (FET) on the input would certainly not have zero input current and infinite input resistance, but a current of <10 pA and an $R_{in} = 10^{12}\ \Omega$ is obtainable and is a reasonable approximation to the ideal conditions. Further, although a CMRR and PSRR of infinity are not possible, there are several commercial op-amps available with values of 140 dB (i.e., a ratio of 10^7). Open loop gains of several precision op-amps now have reached values of $>10^7$, although certainly not infinity. The two most difficult ideal conditions to approach are the ability to handle large output currents and the requirement of a gain independence with frequency.

Using the ideal model conditions, it is quite simple to evaluate the two basic op-amp circuit configurations, (1) the inverting amplifier and (2) the non-inverting amplifier, as designated in Fig. 1-7.

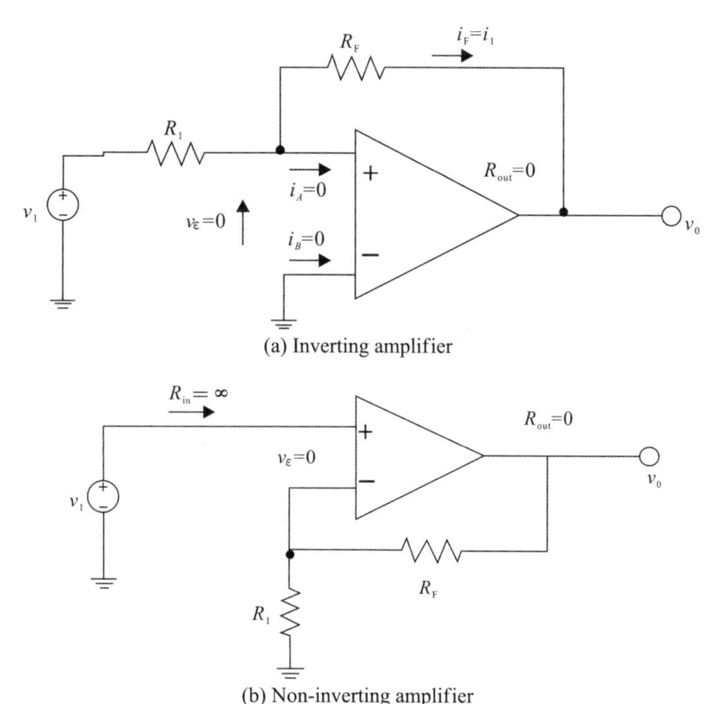

Fig. 1-7 Illustration of the inverting amplifier and the non-inverting amplifier

(Source: E. J. Kennedy, *Operational Amplifier Circuits, Theory and Applications*, New York: Holt, Rinehart and Winston, 1988, pp. 4, 6. With permission.)

For the ideal inverting amplifier, since the open loop gain is infinite and since the output voltage v_0 is finite, then the input differential voltage (often referred to as the error signal) v_ε must approach zero, or the input current is

$$i_1 = \frac{v_1 - v_\varepsilon}{R_1} = \frac{v_1 - 0}{R_1} \tag{1-14}$$

The feedback current i_F must equal i_1 and the output voltage must then be due to the voltage drop across R_F, or

$$v_0 = -i_F R_F + v_\varepsilon = -i_1 R_F = -\left(\frac{R_F}{R_1}\right) v_1 \tag{1-15}$$

The inverting connection thus has a voltage gain v_O/v_1 of $-R_F/R_1$, an input resistance seen by v_1 of R_1 ohms (from Eq. (1-15)), and an output resistance of 0 Ω. By a similar analysis for the non-inverting circuit of Fig. 1-7 (b), since v_ε is zero, then signal v_1 must appear across resistor R_1 producing a current of v_1/R_1 which must flow through resistor R_F. Hence the output voltage is the sum of the voltage drops across R_F and R_1 or

$$v_0 = R_F\left(\frac{v_1}{R_1}\right) + v_1 = \left(1 + \frac{R_F}{R_1}\right) v_1 \tag{1-16}$$

As opposed to the inverting connection, the input resistance seen by the source v_1 is now equal to an infinite resistance, since R_{in} for the ideal op-amp is infinite.

Specialized English Words

operational amplifier　运算放大器
transistor　晶体管
signal processing　信号处理
offset　偏置
inverting　反相
gain　增益
bandwidth　带宽
truth table　真值表
integrated circuit (IC)　集成电路
digital　数字的
analog　模拟的
non-inverting　同相
differential signal　差动信号
frequency　频率
field effect transistor (FET)　场效应晶体管
Boolean algebra　布尔代数

Unit 5 Logic Circuit

5.1 Logic Signals

First let us recap the essential ideas of logic signals. Two-valued signals are used throughout a digital system to represent ON/OFF control actions and to represent the digits of binary numbers. A voltage level will represent each of the two logic values. We might picture a digital system as shown in Fig. 1-8. The logic signals may change over time. It was mentioned that +5 V could be used to represent a logic 1 and 0 V used to represent a logic 0. Using the higher voltage to represent a logic 1 and lower voltage to represent a logic 0 is positive logic representation. Using the higher voltage to represent a logic 0 and the lower voltage to represent a logic 1 is negative logic representation. Often only one representation is used within a system, normally positive logic representation.

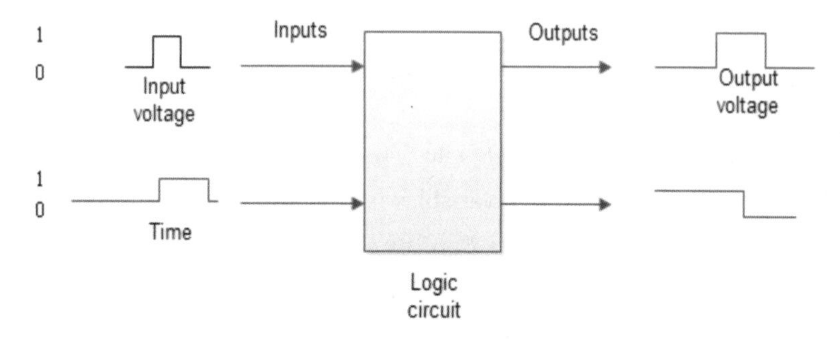

Fig. 1-8 Logic circuit accepting logic voltages and generating logic voltages

5.2 Basic Logic Functions

There are three fundamental operations in Boolean algebra from which all logic functions can be developed, namely: AND , OR and NOT.

These and some other simple functions are implemented by circuits called gates. A gate accepts one or more logic signals and produces a logic output according to a basic logic function of the inputs. Now let us consider the operations more formally.

1. The AND Gate

The AND gate is one of the basic gates that can be combined to form any logic function.

An AND gate can have two or more inputs and performs what is known as logic multiplication.

After completing this section, you should be able to

① Identify an AND gate by its distinctive shape symbol or by its rectangular outline symbol.

② Describe the operation of an AND gate.

③ Generate the truth table for an AND gate with any number of inputs.

(1) AND Gate Truth Table

The logic operation of a gate can be expressed with a truth table that lists all input combinations with corresponding outputs, as illustrated in Tab.1-1 for a 2-input AND gate. The truth table can be expanded to any number of inputs. Although the terms HIGH and LOW tend to give a "physical" sense to the input and output states, the truth table is shown with 1 and 0; a HIGH is equivalent to a 1 and a LOW is equivalent to a 0 in positive logic. For any AND gate, regardless of the number of inputs, the output is HIGH only when all inputs are HIGH.

Tab. 1-1 Truth table for a 2-input AND gate

INPUT		OUTPUT
A	*B*	*X*
0	0	0
0	1	0
1	0	0
1	1	1

1=HIGH, 0=LOW

(2) Logic Expressions for an AND Gate

The logical AND function of two variables is represented mathematically either by placing a dot between the two variables, as $A \bullet B$, or by simply writing the adjacent letters as AB. We will normally use the latter notation because it is easier to write.

Boolean multiplication follows the same basic rules governing binary multiplication, which are as follows:

$$0 \cdot 0 = 0$$
$$0 \cdot 1 = 0$$
$$1 \cdot 0 = 0$$
$$1 \cdot 1 = 1$$

2. The OR Gate

The OR gate is another of the basic gates from which all logic functions are constructed. An OR gate can have two or more inputs and performs what is known as logical addition.

After completing this section, you should be able to

① Identify an OR gate by its distinctive shape symbol or by its rectangular outline symbol.

② Describe the operation of an OR gate.

③ Generate the truth table for an OR gate with any number of inputs.

④ Produce a timing diagram for an OR gate with any specified input wave form.

⑤ Write the logic expression for an OR gate with any number of inputs.

⑥ Discuss an OR gate application.

(1) OR Gate Truth Table

The operation of a 2-input OR gate is described in Tab.1-2. This truth table can be expanded for any number of inputs; but regardless of the number of inputs, the output is HIGH when one or more of the inputs are HIGH.

Tab. 1-2 Truth table for a 2-input OR gate

INPUT		OUTPUT
A	B	X
0	0	0
0	1	1
1	0	1
1	1	1

1=HIGH, 0=LOW

(2) Logic Expressions for an OR Gate

The logical OR function of two variables is represented mathematically by a "+" between the two variables, for example, $A + B$. The plus sign is read as "OR".

Addition in Boolean algebra involves variables whose values are either binary 1 or binary 0. The basic rules for Boolean addition are as follows:

$$0 + 0 = 0$$

$$0 + 1 = 1$$

$$1 + 0 = 0$$

$$1 + 1 = 1$$

3. NOT Gate

A logic variable can be either TRUE or FALSE (a 1 or a 0). We often need to be able to change a variable from one value to the other, i.e., from a 1 to a 0 or from a 0 to a 1. This is the fundamental NOT operation, which has the following definition.

Definition: The NOT operation is applied to a single variable, A, and produces the opposite logic value to A. If A is a 1, NOT A is a 0 and if A is a 0, NOT A is a 1.

The NOT operation on a variable A is written as $\bar{A}$. Some basic electronic circuits to implement gates such as a gate to produce the NOT operation will be developed later. However, it is more convenient in logic design to hide the actual circuit implementations and draw circuits using symbols for the gates. Fig. 1-9 shows the symbol for the NOT gate and its truth table. A truth table lists the output for each possible input value. Since in this case there is only one input, there are two possible input values and it is very reasonable to list the output for each value using a truth table. The NOT gate is sometimes called an inverter, and we talk of inverting a variable, or forming its complement. The bubble on the output of the symbol shows inversion. Bubbles are used freely in logic symbols to show inversion and, as we shall see, can be applied to both inputs and outputs.

Symbol	Symbol
A	$\bar{A}$
0	1
1	0

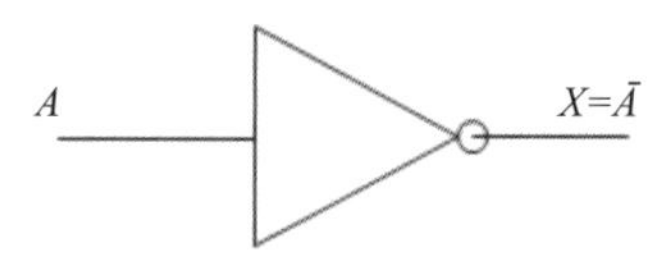

Fig.1-9 NOT gate symbol and truth table

Specialized English Words

logic circuit　逻辑电路
logic level　逻辑电平
positive　正的
negative　负的
logical function　逻辑函数
truth table　真值表
binary　二进制
variable　变量
digital system　数字系统
binary number　二进制数
inverter　反相器
logic representation　逻辑表示法
electronic circuit　电子电路

Unit 6　专业英语简介

6.1 专业英语词汇特点

（1）专业词汇的意义比较专一、固定。科技文献中往往涉及大量的专业术语。不同于普通英语，这些专业术语的词义固定，用来表达确切的含义，如 diode（二极管），resistance（电阻），amplifier（放大器）。需要注意的是，有些词汇除了具有本专业的含义外，在日常英语中还具有其他的含义，如 value 在自动化专业中往往译为（数）值，而在日常英语中往往译为价值。此外，有些词汇在不同学科、不同专业、不同场合的词义是不同的，如 power 在数学专业译为幂、乘方，而在电气工程专业可以译为功率或电力，其中，具体译为功率还是电力，还要视具体场合而定。

（2）广泛使用缩略语、复合词和派生词。科技文献中常常使用缩略语、复合词和派生词。缩略语指的是用一个单词或词组的简写形式（首字母拼接或主干部分）来代表一个完整的形式。例如，AC（交流电）就是 alternate current 的词汇缩写。复合词顾名思义就是将两个或两个以上的单词拼接在一起。通常采用连字符“-”连接两个单词，例如 open-loop（开环）。派生词就是利用前缀和后缀，在原有单词的基础上派生出的单词。例如：layer（层）加上前缀 multi-（多），派生组成 multilayer（多层）；transform（变换）加上后缀 er，派生组成 transformer（变压器）。不难发现，这些前缀和后缀派生出的新的词义是有规律可循的。因此，掌握一些常见的词缀对专业英语的学习和使用是十分必要的。

（3）专业词汇出现比重低。不难发现，在科技文献中，相较于非专业词汇，专业词汇的比重较低。因此，要读懂一篇科技文章，我们不仅要掌握必要的专业词汇，还要注重基础英语词汇的积累。

6.2 专业英语语法特点

由于专业英语所表达的内容具有客观性、准确性、逻辑性、严密性和简明性，专业英语在语法上相应地具备以下几个特点。

（1）复杂长句多。科技语言具有准确、逻辑严密的特点。例如，在描述某一技术过程或操作对象时，专业英语需严格、精确地描述该过程或对象，这样才能客观、准确地传达研究内容。这些要求反映到语法上，就决定了专业英语中多使用长句这一特点。长句中往往包含多个分句和语法层次。这也是学习和掌握专业英语的难点之一。例如：

Conceptually, we can understand resistance if we think about the moving electrons that make up electric current interacting with and being resisted by the atomic structure of the material through which they are moving.

这个句子的主句是“we can understand resistance”，从句是一个由 if 引导的条件状语从句。这个条件状语从句中，又包含了一个 that 引导的定语从句和 which 引导的定语从句。that 引导的从句修饰中心词 electrons，而 which 引导的从句修饰中心词 material。通过语法的分析，我们发现，虽然该句子结构复杂，但是通过运用多个限制性的定语从句，该句生动、准确地描述了什么是电阻。

（2）被动语态多。在专业英语中，为了强调叙述的客观性，句子的主体往往不是动作的发出者，或者是无生命的对象，句子的重点往往不在于动作的执行者，而在于客观存在的事物。有统计认为，在一般的专业英语文献中，至少有 1/3 的动词使用被动语态。例如：

Two-valued signals are used throughout a digital system to represent ON/OFF control actions and to represent the digits of binary numbers.

这个句子采用了被动语态 are used，从而强调了运用二值化信号来表示开关控制动作和二进制数，而并未指出动作的执行者。显然，读者更关心的是用来表示开关控制动作和二进制数的对象，而并不是执行这个动作的人。

（3）广泛使用非谓语动词。相比普通英语，专业英语中使用非谓语动词（动名词、分词、动词不定式）的频率较高。由于专业英语的简明性，我们常常希望用尽可能少的单词来清晰地表达原意，并且当原句中包含多个动作时，为了遵守语法规则，只能有一个谓语动词，其他的动词只能采用非谓语形式。例如：

The resulting circuit may be quite complex, leading to rather laborious calculations.

为了精炼地表达原意，这个句子实际上是将两个句子放到一个句子中，并强化了两句之间的因果关系。这个句子的谓语是 be，另一个动作就用动名词形式 leading。

（4）常使用名词做前置定语，名词或形容词短语作后置定语。这些特点都反映了专业英语的客观性、精炼性、准确性。

6.3 专业英语修辞特点

（1）使用时态较少。普通英语有 16 种时态，但在专业英语中多使用一般现在时、一般过去时和一般将来时 3 种时态。此外，在描述已有的研究背景时，也会用现在完成时。

（2）修辞手法简单。专业英语中很少使用文学英语的修辞手法（如夸张、明喻、隐喻、借喻、拟人、对照）。为了客观、准确、逻辑严密地描述事物或过程，专业英语往往借助图表、公式、流程图、伪代码等方式，以增加科技文献的可读性。

（3）广泛使用逻辑语法词。专业英语注重客观事实论述及逻辑推理。因此，在专业英语中，普遍使用逻辑语法词。例如，表示因果的 because、since、thus、therefore、as a result of、hence，表示转折的 but、however、whereas、while、yet、otherwise，表示并列的 and、as well as、not only ... but also 等。

Unit 7　文献检索方法介绍

7.1 专业文献分类

1. 图书（book）

定义：指主题突出、内容比较系统、知识比较成熟、有完整定型的装帧形式的出版物。

格式：主要责任者. 书名 [M]. 版本（第 1 版不标注）. 出版地：出版者，出版年 .

【例】王兆安，刘进军. 电力电子技术 [M]. 5 版. 北京：机械工业出版社，2013.

2. 期刊（journal）

定义：一种定期或不定期连续出版，通常有固定刊名和年卷期号的出版物。可以按不同的角度分类。若按学术地位，可分为核心期刊和非核心期刊两大类。

格式：主要责任者. 题（篇）名 [J]. 刊名，出版年，卷号（期号）：引文所在的起始或起止页码.

【例】汪晓东，张晨婧仔. “翻转课堂”在大学教学中的应用研究——以教育技术学专业英语课程为例 [J]. 现代教育技术，2013，23(8):11-16.

3. 科技报告（report、technical report）

定义：通常是科学技术工作者围绕某个课题研究所取得的科技成果的总结或对研究进展过程的实际记录。往往涉及高、精、尖科学研究和技术设计及其阶段进展情况和经验教训。

格式：主要责任者. 题名 [R]. 出版地：出版者，出版年：页码.

【例】朱家荷，韩调. 铁路区间通过能力计算方法的研究 [R]. 北京：铁道部科学研究院运输及经济研究所，1989:34.

4. 会议文献（conference paper）

定义：指在各种学术、专题会议上交流、发表的论文、报告和相关文献。会议文献多数以会议录的形式出现，学术性很强，通常代表了某一学科或专业领域的最新研究成果。

格式：析出文献主要责任者. 题（篇）名 [C]// 会议录、论文集主要责任者. 会议录、论文集名. 出版地：出版者，出版年：引文所在起始或起止页码。

【例】朱同春. 浅谈配电网中性点接地方式 [C]// 中国电机工程学会. 全国电网中性点接地方式与接地技术研讨会论文集. 北京：中国电机工程学会，2005:5.

5. 专利文献（patent）

定义：由政府机关或者代表若干国家的区域性组织根据申请而颁发的一种文件，通常是一份详细说明发明的目的、构成及效果的书面技术文件。包括专利说明书、专利公报、专利检索工具以及与专利有关的一切资料。

格式：专利申请者或所有者. 题名 [P]. 专利国别：专利号，公告或公开日期.

【例】黄山松，黄雪松. 一种多功能英语教学管理装置及其教学方法 [P]. 中国专利：CN107194836A，2017-09-22.

6. 学位论文（dissertation、thesis）

定义：高等学校或研究院所的学士、硕士或博士为获得学位，在毕业时撰写的学术性研究论文。通常论述较为系统、详尽，并具有一定的独创性。

格式：主要责任者. 题（篇）名 [D]. 毕业学校所在地：毕业学校，年份.

【例】陈艳君. 中国非英语专业大学生外语听力焦虑研究 [D]. 长沙：湖南师范大学，2005.

7.2 外文文献检索工具

常用外文文献检索工具大致分为两类：一是搜索引擎，如百度学术、谷歌学术，利用庞大的网络资源检索文献；二是国外文献检索系统，如 IEL、Web of Science、Elsevier SDOL 等数据库。

1. Elsevier SDOL

Elsevier SDOL（Elsevier Science Direct OnLine）电子期刊由荷兰爱思唯尔（Elsevier）出版集团出版，已有 180 多年的历史。其出版的期刊是世界上公认的高质量学术期刊。目前该期刊全文库包括 1995 年以来 Elsevier 出版集团旗下各出版社（包括 Academic Press 等）出版的期刊 2200 余种，其中绝大部分为学科核心期刊。被 SCI/SSCI/A&HCI 收录的期刊共 1473 种，占全文库期刊总数的 65%。

覆盖的学科领域有物理学与工程、生命科学、健康科学、社会科学、人文科学、农业与生物、化学及化学工业、医学、计算机、地球科学、工程能源与技术、材料科学、数学、天文、经济、管理等。

2. IEEE/IEE Electronic Library

IEEE/IEE Electronic Library（IEL）数据库提供美国电气电子工程师学会（Institute of Electrical and Electronics Engineers，IEEE）和英国电气工程师学会（Institute of Electrical Engineers, IEE) 出版的 219 种期刊、7151 种会议录、1590 种标准的全文信息。收录了当今世界在电气工程、通信工程和计算机科学领域中近 1/3 的文献，学科领域涉及电气电子工程、计算机科学、人工智能、机器人、自动化控制、遥感和核工程等。

3. SpringerLink

德国 Springer 出版公司是世界著名的出版公司，每年出版期刊超过 2000 种。SpringerLink 是 Springer 出版公司提供其学术期刊及电子图书的在线服务平台，是全球最大的在线科学、技术和医学领域学术资源平台。SpringerLink 收录近 500 种学术期刊，内容涵盖数学、物理和天文学、化学、材料科学、生物医学、生命科学、工程技术、计算机科学、环境科学、地理及经济、法律等学科。

4. ACM Digital Library

ACM Digital Library 是由美国计算机协会（Association for Computing Machinery）创建的提供电子数据服务的全文数据库，收录了该协会的各种电子刊物，包括期刊、会议录、专业杂志、实时通讯及快报等文献，旨在为专业和非专业人士提供了解计算机和信息技术领域资源的窗口。

5. Web of Science

Web of Science 是全球最大、覆盖学科最多的综合性学术资源平台，是有世界影响力的多学科学术文献文摘索引数据库。它包括三个主要的引文数据库，即科学引文索引（Science Citation Index，SCI）、社会科学引文索引（Social Sciences Citation Index，SSCI）、艺术与人文科学索引（Arts & Humanities Citation Index，A&HCI）以及两个化学信息事实型数据库（CCR 和 IC）和科学引文检索扩展版、科技会议文献引文索引和社会科学以及人文科学会议文献引文索引三个引文数据库。收录了自然科学、工程技术、生物医学等各个研究领域最具影响力的 8700 多种核心学术期刊。利用 Web of Science 丰富而强大的检索功能，可以访问最为可靠且涉及多个学科的整合科研成果，从而全面了解有关某一学科、某一课题的研究信息。

7.3 检索关键词选取

关键词是从文章的题名、摘要、正文中抽出的具有实在意义的单词或术语，是具有实质意义、最能表达全文内容主题的词或词组。关键词的确定方法包括：（1）分析课题，提取概念，选择能准确反映检索意义的关键词；（2）整理概念，转化关键词，扩充同义词汇，同一主题概念下的同义词、相关词、上位词和下位词都可作为检索词；（3）运用上述分析所得的检索词，在所选择的检索工具中进行试检索，根据检索结果进而确定适合课题的关键词。还可以利用检索结果中提供的同义词链，选择其中的一个关键词进行检索，全面地检索到所需信息。

选取关键词时还需注意以下几点：（1）分析课题的内容实质和检索目的；（2）选用规范词，最好选择专业性强的词，少用自创词；（3）排除无关概念和不易检索的检索词，不选择介词、副词、连词、助词等；（4）分析潜在的主题概念，找同义词、近义词和表达同一概念词的不同词汇。

7.4 影响因子

影响因子（Impact factor，IF）是一个国际上通行的衡量学术期刊影响力的重要指标。具体计算方法是该期刊前两年发表的论文在当年的被引总次数与该期刊前两年发表的论文总数之比。一般期刊的影响因子都是由美国 Thomson Reuters 发布的期刊引证报告（Journal Citation Reports，JCR）统计并计算的。不是所有的期刊都能查到影响因子。对于一些综合类或者研究领域较广的期刊，其引用率往往也比较高，也就比较容易有较高的影响力。值得说明的是，影响因子并非最客观的评价期刊影响力的标准。

实训课程一：文献检索

1. 快速检索——与某个主题相关的文献

选择一个电气或自动化专业研究课题，检索 3 篇参考文献（期刊论文或会议论文），可通过以下英文数据库检索：

（1）Elsevier SDOL（http://www.sciencedirect.com/）。

（2）IEEE/IEE Electronic Library（http://ieeexplore.ieee.org/Xplore/home.jsp）。

（推荐 http://digital-library.theiet.org/）。

（3）SpringerLink（http://link.springer.com/）。

实例：

Step 1：选择课题。

"工业机器人的运动控制轨迹规划及控制策略研究"。

Step 2：选择合适的关键字。

（industrial）robot，motion control，path planning。

Step 3：选择数据库。

IEL（http://ieeexplore.ieee.org/Xplore/home.jsp）。

Step 4：通过期刊影响因子或引用次数选择合适的文章。

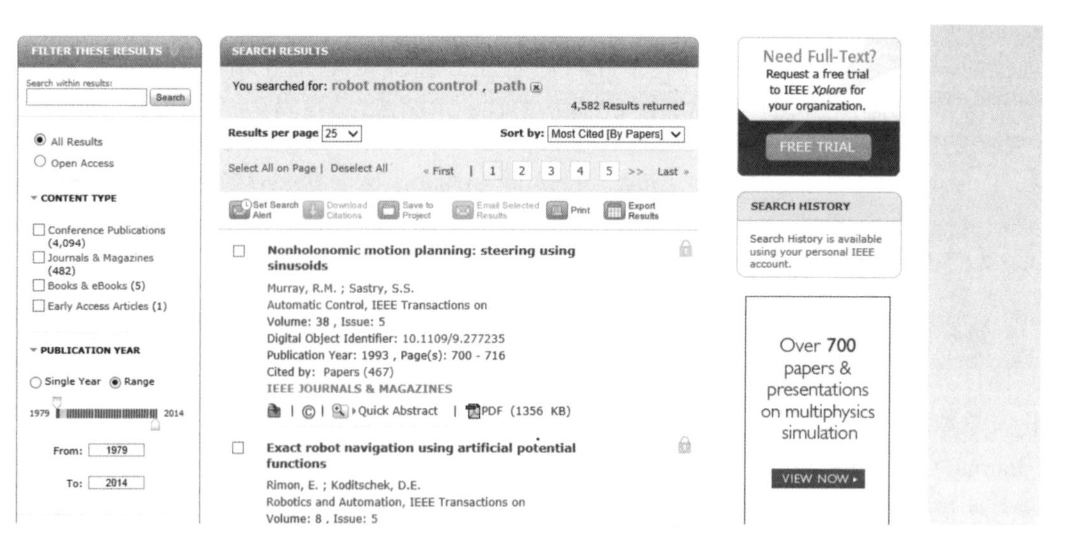

2. 快速检索——查找已知文献

实例：检索某篇文献（已知标题或者所发表期刊来源）

Title: Smart Grid—The New and Improved Power Grid: A Survey.

Journal: *IEEE Communications Surveys & Tutorials.*

Chapter 2

ELECTRIC MACHINE

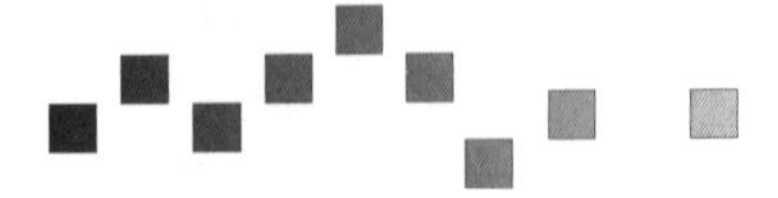

Unit 1 Magnetic Field Theory

1.1 Introduction

Magnetic fields are the fundamental mechanism by which energy is converted from one form to another in motors, generators, and transformers. Four basic principles describe how magnetic fields are used in these devices:

(1) Current-carrying wire produces a magnetic field in the area around it.

(2) A time-changing magnetic field induces a voltage in a coil of wire if it passes through that coil. (This is the basis of transformer action.)

(3) A current-carrying wire in the presence of a magnetic field has a force induced on it. (This is the basis of motor action.)

(4) A moving wire in the presence of a magnetic field has a voltage induced in it. (This is the basis of generator action.)

1.2 Production of a Magnetic Field

The basic law governing the production of a magnetic field by a current is Ampere's law:

$$\oint H \cdot \mathrm{d}l = I_{net} \tag{2-1}$$

where H is the magnetic field intensity produced by the current I_{net}, and $\mathrm{d}l$ is a differential element of length along the path of integration. In SI units, I is measured in amperes and H is measured in ampere-turns per meter. To better understand the meaning of this equation, it is helpful to apply it to the simple example in Fig. 2-1. Fig. 2-1 shows a rectangular core with a winding of N turns of wire wrapped about one leg of the core. If the core is composed of iron or certain other similar metals (collectively called ferromagnetic materials), essentially all the magnetic field produced by the current will remain inside the core, so the path of integration in Ampere's law is the mean path length of the l_c. The current passing within the path of integration I_{net} is then Ni, since the coil of wire cuts the path of integration N times while carrying current i. Ampere's law thus becomes

$$Hl_c = Ni \tag{2-2}$$

where H is the magnitude of the magnetic field intensity vector H. Therefore, the magnitude of the magnetic field intensity in the core due to the applied current is

$$H = \frac{Ni}{l_c} \tag{2-3}$$

The magnetic field intensity H is in a sense a measure of the "effort" that a current is putting to the establishment of a magnetic field.

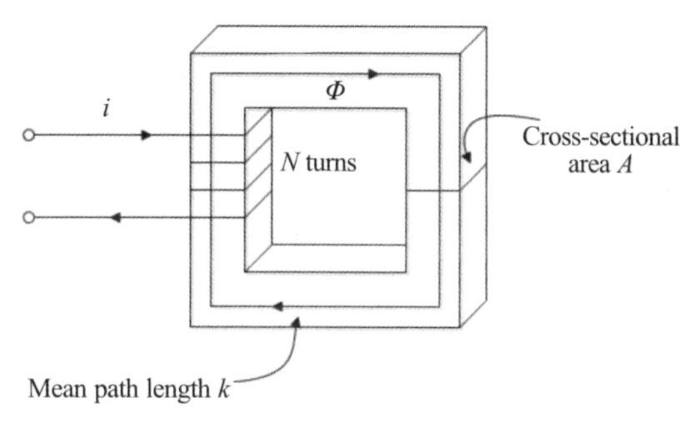

Fig. 2-1 A simple magnetic core

1.3 Faraday's Law of Electromagnetic Induction

It is now time to examine the various ways in which an existing magnetic field affects its surroundings. The first major effect to be considered is called *Faraday's law*. The operation of transformers and generators is on the basis of it. Faraday's law states that if a flux passes through a coil of N turns, a voltage, which is directly proportional to the rate of change in the flux with respect to time, can be induced in the coil. It can be expressed in equation

$$e=-N\frac{d\phi}{dt} \tag{2-4}$$

where e is the voltage induced in the coil; N is the number of turns of the coil; Φ is the flux passing through the coil.

The minus sign in the equation is given by *Lenz's law*. Lenz's law states that the direction of the voltage builds up in the coil in a way that if the coil ends are short circuited, it will produce current that will cause a flux opposing to the change of original flux. Since the induced voltage opposes the change that causes it, a minus sign is added in Eq. (2-4).

If the flux presented in each turn of the coil is not the same, Eq. (2-4) can be rewritten as:

$$e=-\sum_{i=1}^{N}e_i=-\sum_{i=1}^{N}\frac{\mathrm{d}\phi_i}{\mathrm{d}t}=-\frac{\mathrm{d}}{\mathrm{d}t}(\sum_{i=1}^{N}\phi_i)=-\frac{\mathrm{d}\Psi}{\mathrm{d}t} \tag{2-5}$$

where Ψ is the flux linkage of the coil, its unit is Weber-turn, and

$$\psi=\sum_{i=1}^{N}\phi_i \tag{2-6}$$

When a conductor with proper orientation moves through a magnetic field, a voltage will be induced in it. When movement direction of the conductor and direction of the field are perpendicular to each other, the voltage induced in the wire is in direction perpendicular to the movement and field direction. Magnitude of the voltage can be calculated by

$$e=Blv \tag{2-7}$$

where B is the flux density of the magnetic field [T]; l is the active length of the conductor in the magnetic field [m]; v is the relative speed of the conductor [m/s].

The direction of induced voltage can be determined by *right hand rule*. If the direction of the flux density B goes through the palm of right hand vertically and the thumb of the right hand points at the direction of the conductor moving, the direction of the voltage induced in the conductor is shown by where the middle finger of the right hand is pointing at.

1.4 Production of Induction Force on a Wire

The second effect of a magnetic field on its surroundings is that it induces a force on a current-carrying wire within the field. The basic concept involved is illustrated in Fig. 2-2 (omitted). The figure shows a uniform magnetic field of flux density B with a conductor in it. The conductor is l meters long and it contains a current of i amperes. The force induced on the conductor is given by

$$f=Bli \tag{2-8}$$

where B is the flux density of the magnetic field [T]; l is the active length of the conductor in the magnetic field [m]; i is the current of the conductor [A].

The direction of the force is given by *left-hand rule* (Fig. 2-3). If the direction of the flux density B goes through the palm of left hand vertically and the index finger of the left hand points at the direction of the current in the conductor, the direction of the resultant force on the conductor is the same as the direction that the thumb is

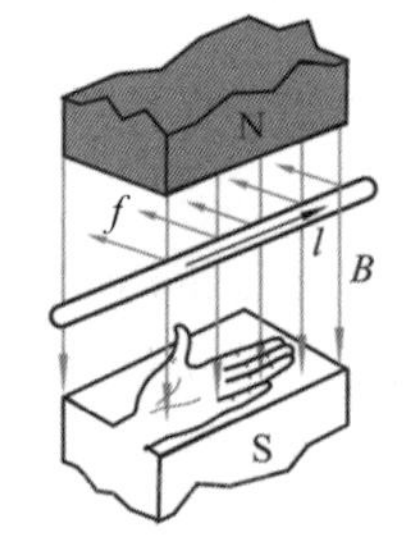

Fig. 2-3 Left-hand rule

pointing at.

1.5 Induced Voltage on a Conductor Moving in a Magnetic Field

There is a third major way in which a magnetic field interacts with its surroundings. If a wire with the proper orientation moves through a magnetic field, a voltage is induced in it. The voltage induced in the wire is given by

$$e_{ind} = (v \times B) \cdot l \tag{2-9}$$

where

v = velocity of the wire

B = magnetic flux density vector

l = length of conductor in the magnetic field

Vector l points along the direction of the wire toward the end making the smallest angle with respect to the vector $v \times B$. The voltage in the wire will be built up so that the positive end is in the direction of the vector $v \times B$.

Specialized English Words

motor 马达；电动机

transformer 变压器

coil 线圈

core 磁芯

electromagnetic induction 电磁感应

flux linkage 磁链

generator 产生器；发电机

magnetic field 磁场

magnetic field intensity 磁场强度

ferromagnetic 铁磁性

flux 通量

magnetic flux density 磁通量密度

Unit 2　DC Machines

2.1 Introduction

DC machines are electromechanical energy-conversion devices. They can be operated either as generators or motors (because mechanical energy and electrical energy conversion is reversible with a DC machine). When a DC machine is operated as a generator, mechanical energy will be converted to DC electrical energy. When it is used as a motor, electrical energy will be converted to mechanical energy. In principle, DC machines are similar to AC machines, since both DC and AC machines have AC voltages and currents within their windings. However, only DC machines have DC outputs because of their unique commutation mechanism commutator.

The physical structure of the machine consists of two parts: the *stator* or stationary part and the *rotor* or rotating part. The stationary part of the machine consists of the *frame*, which provides physical support, and the *pole pieces*, which project inward and provide a path for the magnetic flux in the machine. The ends of the pole pieces that are near the rotor spread out over the rotor surface to distribute its flux evenly over the rotor surface. These ends are called the *pole shoes*. The exposed surface of a pole shoe is called a *pole face,* and the distance between the pole face and the rotor is called the *air gap*.

There are two principal windings on a DC machine: the *armature windings* and the *field windings*. The armature windings are defined as the windings in which a voltage is induced, and the field windings are defined as the windings that produce the main magnetic flux in the machine. In a normal DC machine, the armature windings are located on the rotor, and the field windings are located on the stator. Because the armature windings are located on the rotor, a DC machine's rotor itself is sometimes called an armature.

2.2 Pole and Frame Construction

The main poles of older DC machines were often made of a single cast piece of metal, with the field windings wrapped around it. They often had bolted-on laminated tips to reduce core losses in the pole faces. Since solid-state drive packages have become common, the main poles of newer machines are made entirely of laminated material. This is true because there is a much higher AC content in the power supplied to DC motors driven by solid-state drive packages, resulting in much higher eddy current losses in the stators of the machines. The pole faces are typically either chamfered or eccentric in construction, meaning that the outer tips of a pole face are spaced slightly further from the rotor's surface than the center of the pole face is. This action increases the reluctance at the tips of a pole face and therefore reduces the flux-bunching effect of armature reaction on the machine.

The poles on DC machines are called *salient poles*, because they stick out from the surface of the stator.

The interpoles in DC machines are located between the main poles. They are more and more commonly of laminated construction, because of the same loss problems that occur in the main poles.

2.3 Rotor or Armature Construction

The rotor or armature of a DC machine consists of a shaft machined from a steel bar with a core built up over it. The core is composed of many laminations stamped from a steel plate, with notches along its outer surface to hold the armature windings. The commutator is built onto the shaft of the rotor at one end of the core. The armature coils are laid into the slots on the core, and their ends are connected to the commutator segments.

2.4 Commutator and Brushes

The commutator in a DC machine is typically made of copper bars insulated by a mica-type material. The copper bars are made sufficiently thick to permit normal wear over the lifetime of the motor. The mica insulation between commutator segments is harder than the commutator material itself, so as a machine ages, it is often necessary to undercut the commutator insulation to ensure that it does not stick up above the level of the copper bars.

The brushes of the machine are made of carbon, graphite, metal graphite, or a mixture of carbon and graphite. They have a high conductivity to reduce electrical losses and a low coefficient of friction to reduce excessive wear. They are deliberately made of much softer material than that of the commutator segments, so that the commutator surface will experience very little wear. The choice of brush hardness is a compromise: If the brushes are too soft, they will have to be replaced too often; but if they are too hard, the commutator surface will wear excessively over the life of the machine.

All the wear that occurs on the commutator surface is a direct result of the fact that the brushes must rub over them to convert the AC voltage in the rotor wires to DC voltage at the machine's terminals. If the pressure of the brushes is too great, both the brushes and the commutator bars wear excessively. However, if the brush pressure is too small, the brushes tend to jump slightly and a great deal of sparking occurs at the brush-commutator segment interface. This sparking is equally bad for the brushes and the commutator surface. Therefore, the brush pressure on the commutator surface must be carefully adjusted for maximum life.

Another factor which affects the wear on the brushes and segments in a DC machine commutator is the amount of current flowing in the machine. The brushes normally ride over the commutator surface on a thin oxide layer, which lubricates the motion of the brushes over the segments. However, if the current is very small, that layer breaks down, and the friction between the brushes and the commutator is greatly increased. This increased friction contributes to rapid wear. For maximum brush life, a machine should be at least partially loaded all the time.

2.5 Winding Insulation

Other than the commutator, the most critical part of a DC motor's design is the insulation of its windings. If the insulation of the motor windings breaks down, the motor shorts out. The repair of a machine with shorted insulation is quite expensive, if it is even possible. To prevent the insulation of the machine windings from breaking down as a result of overheating, it is necessary to limit temperature of the windings. This can be partially done by providing a cooling air circulation over them, but ultimately the maximum winding temperature limits the maximum power that can be supplied continuously by the machine.

2.6 Principles of DC Generators

Fig. 2-4 shows an elementary DC generator composed of a coil that revolves between a pair of N, S poles of a permanent magnet at 60 r/min. The rotation is due to an external driving torque from a device such as a motor (not shown). The coil is connected to two slip rings mounted on the shaft. The slip rings are connected to an external load by means of two stationary brushes x and y. Brushes x, y can switch from one slip ring to another every time when the polarity of the field pole to meet is about to change. Brush x will always be positive and brush y will be negative. This configuration can be achieved by using a commutator (Fig. 2-3). A commutator in its simplest form is composed of a slip ring that is cut in half, with each segment insulated from the other as well as from the shaft. One segment is connected to coil end A and the other to coil end D. The commutator revolves with the armature coil and the voltage between segments is picked up by two stationary brushes x and y.

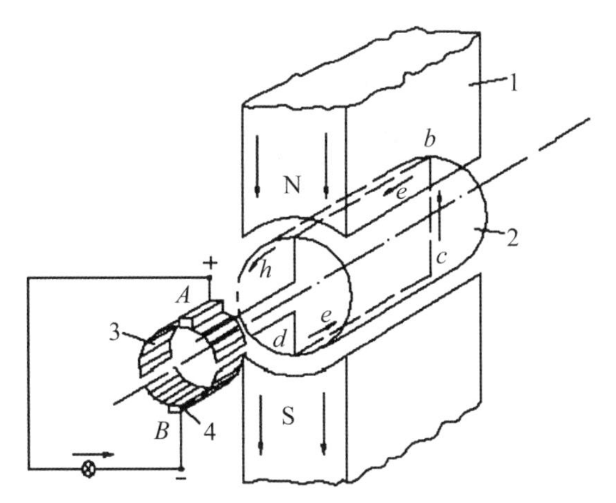

Fig. 2-4 An elementary DC generator

The induced voltage is therefore maximum (20V) when the armature coil lies on the same plane with the magnetic field as shown in Fig. 2-4. No flux is cut when the armature coil is perpendicular to the field, consequently the voltage is zero momentarily. Another feature of the induced voltage in the armature coil is that its polarity changes every time when the coil makes a half turn. The coil in our example revolves at a uniform speed, therefore the voltage can be represented as a function of the angle of rotation. The wave shape depends upon the shape of the N, S poles. We assume that the sinusoidal wave shown below can be generated by the designed poles.

The voltage between brushes x and y pulsates but the voltage between the two brushes never changes the polarity. The alternating voltage in the coil is therefore rectified by the commutator.

Due to the constant polarity of the voltage between brushes, current in the external load always flows in the same direction. The machine shown in Fig. 2-4 is called a direct-current generator or dynamo.

2.7 Principles of DC Motors

Direct-current motors are built in the same way as that of generators, consequently, a DC machine can be operated either as a motor or a generator. Fig. 2-5 shows an elementary DC motor composed of a coil and a pair of N, S poles of permanent magnets. The coil initially at rest is connected to a DC voltage U by means of a switch (Fig. 2-5). A magnetic field is created by the permanent magnets.

As soon as the switch is closed, current will flow into the coil. The coil will immediately be subjected to a force because it is immersed in the magnetic field created by the permanent magnets. The forcc produces a powerful torque, causing the coil to rotate.

From Fig. 2-4 and Fig. 2-5, we can find that the DC generator and the DC motor have the same construction. When the external condition (DC voltage or driving torque) changes, the operational mode of the machine will change correspondingly and this is called electric machine's reversible operation principle.

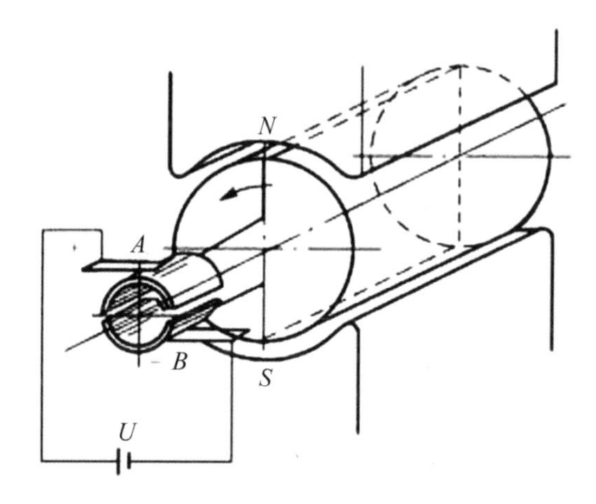

Fig.2-5 Schematic diagram of an elementary DC motor

Specialized English Words

electromechanical 电机的；机电的

direct current（DC） 直流电流

winding 绕组

stator 定子

pole 极

salient pole 凸极

slot 槽口；间隙

brush 刷；刷子

slip ring 滑环

conversion 转换

alternating current (AC) 交流电流

commutator 换向器

rotor 转子

armature 电枢

interpole 换向极

insulation 绝缘

coefficient 系数

permanent magnet 永久磁铁

Unit 3 Transformer

3.1 Construction of Transformers

A transformer is a device that changes AC electric power at one frequency and voltage level to AC electric power at the same frequency and another voltage level through the action of a magnetic field. As shown in Fig. 2-6, it consists of two or more coils of wire wrapped around a common ferromagnetic core. These coils are (usually) not directly connected. The only connection between the coils is the common magnetic flux present within the core.

One of the transformer windings is connected to a source of AC electric power, and the second (and perhaps third) transformer winding supplies electric power to loads. The transformer winding connected to the power source is called the *primary winding* or input winding with N_1 turns, and the winding connected to the loads is called the *secondary winding* or output winding with N_2 turns. If there is a third winding on the transformer, it is called the *tertiary winding*.

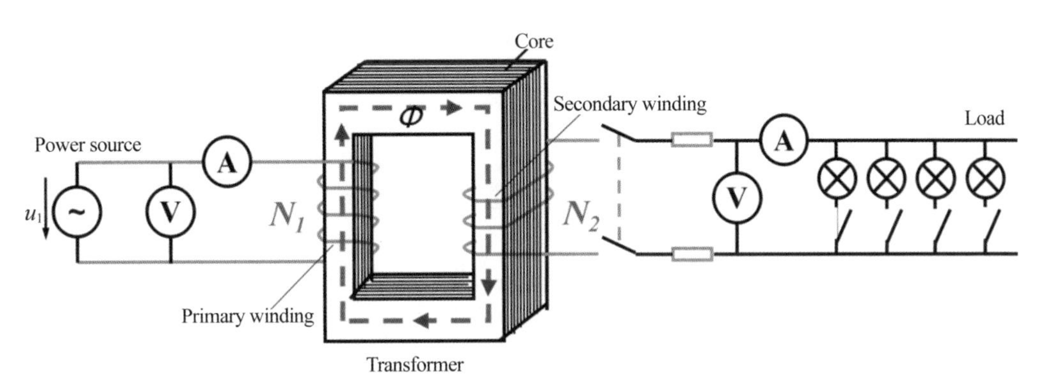

Fig. 2-6 Power transformer with core-form construction

Transformers are also used for other purposes, such as voltage sampling, current sampling and impedance transformation. However, this unit is primarily devoted to the power transformer.

Power transformers are constructed using either the core-form construction or the shell-form construction. The former one consists of a simple rectangular laminated piece of steel with the transformer windings wrapped around two sides of the rectangle. The later construction consists of a three legged laminated core with the windings wrapped around the center leg. In order to minimize hysteresis and eddy current losses, the core is made of silicon steel laminations with high permeability. The thickness of each laminated steel ranges between 0.35 mm to 0.5 mm. All laminations are insulated and separated from each other.

3.2 Ideal Transformer

An ideal transformer is a lossless device with an input winding and an output winding. The relationships between the input voltage and the output voltage, and between the input current and the output current, are given by two simple equations. Fig. 2-7 shows an ideal transformer.

The transformer shown in Fig. 2-7 has N_P turns of wire on its primary side and N_S turns of wire on its secondary side. The relationship between the voltage $v_P(t)$ applied to the primary side of the transformer and the voltage $v_S(t)$ produced on the secondary side is

$$\frac{v_P(t)}{v_S(t)} = \frac{N_P}{N_S} = a \tag{2-10}$$

where a is defined to be the turns ratio of the transformer, as the following:

$$a = \frac{N_P}{N_S} \tag{2-11}$$

The relationship between the current $i_P(t)$ flowing into the primary side of the transformer and the current $i_S(t)$ flowing out of the secondary side of the transformer is

$$N_P i_P(t) = N_S i_S(t) \tag{2-12a}$$

or

$$\frac{i_P(t)}{i_S(t)} = \frac{1}{a} \tag{2-12b}$$

In terms of phasor quantities, these equations are

$$\frac{V_P}{V_S} = a \tag{2-13}$$

and

$$\frac{\quad}{I} \quad \frac{\quad}{a} \tag{2-14}$$

Notice that the phase angle of V_P is the same as the angle of V_S and the phase angle of I_P is the same as the phase angle of I_S. The turns ratio of the ideal transformer affects the magnitudes of the voltages and currents, but not their angles.

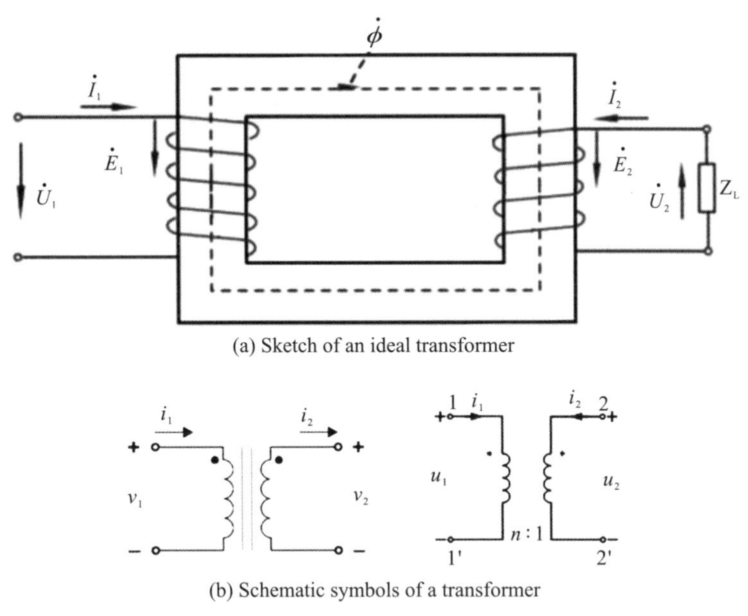

(a) Sketch of an ideal transformer

(b) Schematic symbols of a transformer

Fig. 2-7 Ideal transformer

Fig. 2-7 tells the polarity of the voltage and current on the secondary side of the transformer. The relationship is as follows:

(1) If the primary voltage is positive at the dotted end of the winding with respect to the undotted end, then the secondary voltage will be positive at the dotted end also. Voltage polarities are the same with respect to the dots on each side of the core.

(2) If the primary current of the transformer flows into the dotted end of the primary winding, the secondary current will flow out of the dotted end of the secondary winding.

3.3 Equivalent Circuit of a Transformer

The losses that occur in real transformers have to be accounted for in any accurate model of transformer behavior. The major items to be considered in the construction of such a model are:

(1) Copper (I^2R) losses. Copper losses are the resistive heating losses in the primary and secondary windings of the transformer. They are proportional to the square of the current in the windings.

(2) Eddy current losses. Eddy current losses are resistive heating losses in the core of the transformer. They are proportional to the square of the voltage applied to the transformer.

(3) Hysteresis losses. Hysteresis losses are associated with the rearrangement of the magnetic domains in the core during each half cycle. They are a complex and nonlinear function of the voltage applied to the transformer.

(4) Leakage flux. The fluxes ϕ_{LP} and ϕ_{LS} which escape the core and pass through only one of the transformer windings are leakage fluxes. These escaped fluxes produce a leakage inductance in the primary and secondary coils, and the effects of this inductance must be accounted for.

The easiest effect to model is the copper losses. They are modeled by placing a resistor R_P in the primary circuit of the transformer and a resistor R_S in the secondary circuit.

The leakage flux in the primary windings ϕ_{LP} produces a voltage e_{LP} given by

$$e_{LP}(t) = N_P \frac{\mathrm{d}\phi_{LP}}{\mathrm{d}t} \tag{2-15a}$$

And the leakage flux in the secondary windings produces a voltage e_{LS} given by

$$e_{LS}(t) = N_S \frac{\mathrm{d}\phi_{LS}}{\mathrm{d}t} \tag{2-15b}$$

Since much of the leakage flux path is through air, and since air has a constant reluctance much higher than the core reluctance, the flux ϕ_{LP} is directly proportional to the primary circuit current i_P, and the flux ϕ_{LS} is directly proportional to the secondary current i_S.

$$\phi_{LP} = (\xi N_P) i_P \tag{2-16a}$$

$$\phi_{LS} = (\xi N_S) i_S \tag{2-16b}$$

where

ξ = permeance of the flux path

N_P = number of turns on the primary coil

N_S = number of turns on the secondary coil

Substitute Eq. (2-16) into Eq. (2-15). The result is

$$e_{LP}(t) = N_P \frac{\mathrm{d}}{\mathrm{d}t}(\xi N_P) i_P = N_P^2 \xi \frac{\mathrm{d}i_P}{\mathrm{d}t} \tag{2-17a}$$

$$e_{LS}(t) = N_S \frac{\mathrm{d}}{\mathrm{d}t}(\xi N_S) i_S = N_S^2 \xi \frac{\mathrm{d}i_S}{\mathrm{d}t} \tag{2-17b}$$

The constants in these equations can be lumped together. Then

$$e_{LP}(t) = L_P \frac{\mathrm{d}i_P}{\mathrm{d}t} \tag{2-18a}$$

$$e_{LS}(t) = L_S \frac{\mathrm{d}i_S}{\mathrm{d}t} \tag{2-18b}$$

where $L_P = N_P^2\xi$ is the leakage inductance of the primary coil and $L_S = N_S^2\xi$ is the leakage inductance of the secondary coil. Therefore, the leakage flux will be modeled by primary and secondary inductors.

The magnetization current i_m is a current proportional (in the unsaturated region) to the voltage applied to the core and *lagging the applied voltage by* 90°, so it can be modeled by a reactance X_m connected across the primary voltage source. The core loss current i_{h+e} is a current proportional to the voltage applied to the core that is in phase with the applied voltage, so it can be modeled by a resistance R_c connected across the primary voltage source. (Remember that both these currents are really nonlinear, so the inductance X_m and the resistance R_c are, at best, approximations of the real excitation effects.)

The resulting equivalent circuit is shown in Fig. 2-8. In this circuit, R_P is the resistance of the primary winding, X_P ($=\omega L_P$) is the reactance due to the primary leakage inductance, R_S is the resistance of the secondary winding, and X_S($= \omega L_S$) is the reactance due to the secondary leakage inductance. The excitation branch is modeled by the resistance R_c (hysteresis and core losses) in parallel with the reactance X_m (the magnetization current).

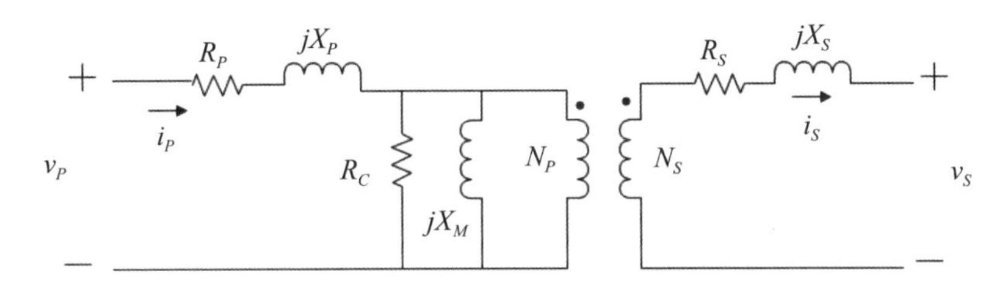

Fig. 2-8 The model of a real transformer

Specialized English Words

transformer 变压器
primary winding 一次绕组
tertiary winding 三次绕组
impedance 阻抗
eddy current 涡流
phasor quantity 相量
positive 正的
nonlinear 非线性的
inductance 电感
permeance 磁导
magnetization 磁化
saturated 饱和的
ferromagnetic 铁磁性
secondary winding 二次绕组
voltage sampling 电压采样
hysteresis 磁滞
permeability 磁导率
polarity 极性
equivalent circuit 等效电路
leakage flux 漏通量
reluctance 磁阻
inductor 电感器
lag 滞后；延迟
reactance 电抗

Unit 4 Synchronous Machine

4.1 Construction of a Synchronous Machine

Commercial synchronous machines are built with either a stationary or a rotating DC magnetic field.

A stationary field synchronous machine has the same outward appearance as a DC machine. The salient poles create a DC field, which is cut by a revolving armature. The armature possesses a 3-phase winding, of which the terminals are connected to three slip rings that are mounted on the shaft. A set of brushes, which are sliding on the slip rings, enables the armature to be connected to the external 3-phase circuit. For a synchronous generator, the armature is driven by a gasoline engine, or some other source of motive power. The stationary-field synchronous generators are used when the power output is less than 5 kV · A.

A revolving-field synchronous machine has a stationary armature, which is called a stator. The 3-phase stator winding is directly connected to the external 3-phase circuit, without going through those large, unreliable slip rings and brushes. The field windings are on the rotor and are excited by a DC voltage. The revolving-field synchronous generators are usually used as the synchronous machines with a high voltage and a large capacity.

According to the shape of the magnetic poles on the rotor, the revolving-field synchronous machines are classified into two types. The first is a non-salient (cylindrical) construction, as shown in Fig. 2-9 (a). The field windings are distributed windings placed in the slots on the surface of the cylindrical rotor, so the air gap of a non-salient pole machine is uniform. This generator is usually called the turbo generator, which uses the steam turbine as the prime mover and has 2 or 4 poles. This generator also operates the best at relatively high speeds. The second is a salient construction (Fig. 2-9 (b)), which is classified as the low speed type (engine-or water-driven). This is characterized physically by having salient poles, a large diameter, and a short axial length. Its air gap is not uniform. The field windings are concentrated windings. This generator is usually called the hydroelectric generator, which has a relatively large number of poles.

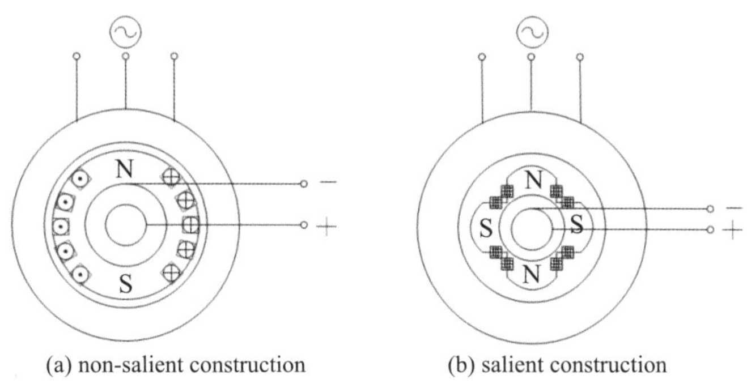

(a) non-salient construction (b) salient construction

Fig. 2-9 Two basic revolving-field synchronous machines

There are two terms used commonly to describe the windings on a machine, one is the *field winding,* the other is the *armature winding.* In general, the first term applies to the windings that produce a main magnetic field in a machine. The latter applies to the windings where the main voltage is induced. For a synchronous machine, the field windings are on the rotor, so the terms rotor windings and field winding are used interchangeably. Similarly, the terms stator windings and armature windings are used interchangeably as well.

As shown in Fig. 2-9, a three-phase synchronous machine has two main parts: a stationary stator and a revolving rotor. The rotor is separated from the stator by a uniform air gap (non-salient pole machine) or a non-uniform air gap (salient pole machine).

4.2 Stator

The stator of a synchronous machine consists of stator core, stator winding, frame and cover. The stator core is built up from the toothed segments of the high-quality silicon-iron steel laminations (0.5 mm thick) and is covered with an insulating varnish. From an electrical standpoint, the stator of a synchronous machine is identical to that of a 3-phase induction motor. It is composed of a cylindrical laminated core, containing a set of slots that carries a 3-phase lap winding. The winding is always Y-connected and the neutral is connected to a ground.

4.3 Rotor

The rotor of a sychronous machine is composed of rotor core, field winding, protecting ring, fan and shaft.

Fig. 2-10 shows the salient pole and the windings on the rotor of a hydroelectric generator. The salient poles of the rotor are composed of iron laminations that are much thicker (1~2 mm). These laminations are not insulated, because the DC flux that they carry does not vary. The 6 round holes in the face of the salient pole carry the bars of a squirrel-cage winding, and all the copper bars are shorted with two end rings.

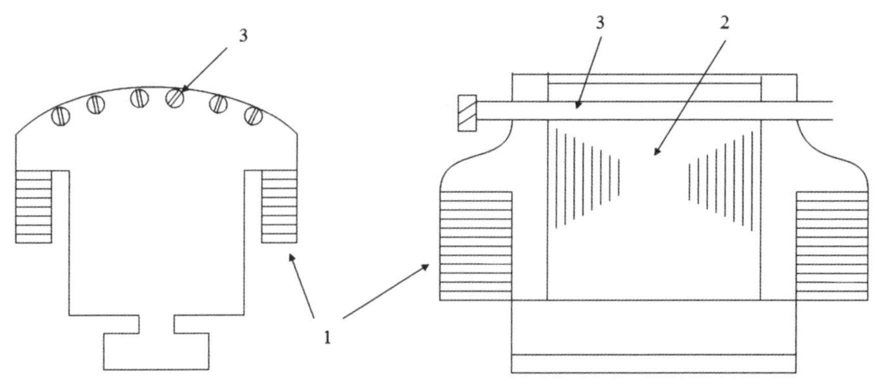

Fig. 2-10 Salient pole of a hydroelectric generator
1-Field winding; 2-Field pole; 3-Damper

Fig. 2-10 shows the salient pole of a hydroelectric generator. The squirrel-cage winding is called a *damper winding* sometimes.

Under normal condition, the damper winding does not carry any current, because the rotor turns at a synchronous speed. However, when the rotor speed begins to fluctuate for a certain reason, a voltage will be induced in the damper winding, causing a large current to flow therein. The current reacts with the magnetic field of the stator, producing a force which damps the oscillation of the rotor. The damper winding also tends to maintain a balanced 3-phase voltages between the lines, even when the line currents are unequal due to the unbalanced load conditions.

4.4 Operation Mode of a Synchronous Machine

When a synchronous machine operates in loaded condition at a steady state, the stator (armature) three-phase windings carry the balanced three-phase currents, producing a stator magnetic field which rotates at synchronous speed. Meanwhile, a DC excitation current is supplied to the field winding on the rotor, producing a rotor magnetic field which revolves at the same speed and in the same direction as the stator magnetic field. The magnetic fields of the stator and the rotor are therefore stationary with respect to each other, producing a resultant air-gap magnetic field which is constant in amplitude and rotates at a synchronous speed.

Power angle δ is defined as the angle between the axis of the rotor magnetic field and the axis of the resultant stator magnetic field (Fig. 2-11).

Notice that if two magnetic fields are present in a machine, then a torque will be generated and it will tend to line up the two magnetic fields. There are two magnetic fields in a synchronous machine, one magnetic field is produced by the stator currents and the other one by the rotor currents. Then, a torque will be induced in the rotor, which will cause the rotor to turn and align itself with the stator magnetic field, and the electromechanical energy conversion can be achieved in the machine.

The slip s is an important variable to reflect operation modes of an induction machine. Similarly, the power angle δ is an important variable to reflect that of a synchronous machine. There are three operation modes of a synchronous machine: the generator, the motor and the compensator.

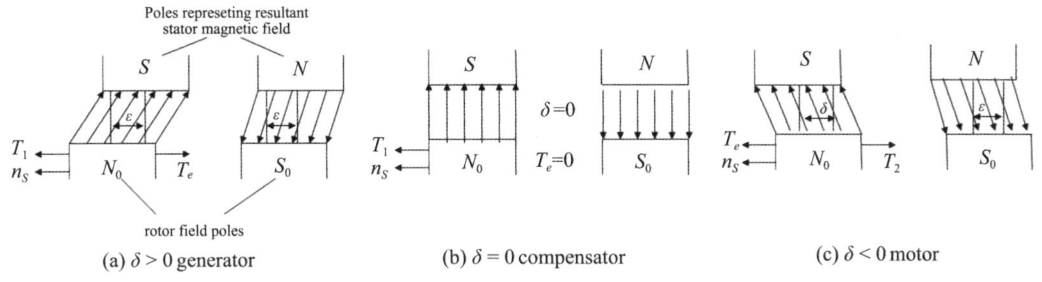

Fig. 2-11 Three operating modes of a synchronous machine.

If the rotor magnetic field is leading the resultant stator magnetic field by δ degree, the power angle δ will be positive. At this condition, a braking electromagnetic torque is produced on the rotor. The prime mover must provide a driving torque to overcome the braking electromagnetic torque, so the mechanical energy is supplied from the prime mover to the rotor. The electric energy is delivered from the stator to the power systems, and this synchronous machine works as a generator (Fig. 2-11 (a)).

If the axis of the rotor magnetic field and the resultant stator magnetic field are aligned, the power angle δ will be zero. At this condition, no electromagnetic torque is produced on the rotor, and then no electromechanical energy conversion will be achieved. But the synchronous machine can absorb or deliver a reactive power in the power systems, and it can regulate the power factor of power systems. It works as a compensator (Fig. 2-11 (b)).

If the rotor magnetic field rotates behind the resultant stator magnetic field, the power angle δ will be negative. At this condition, a driving electromagnetic torque is produced on the rotor. This torque pulls a load that is connected to the shaft. The rotor delivers a mechanical energy to the load, and this synchronous machine works as a motor (Fig. 2-11 (c)).

Specialized English Words

synchronous 同步的
armature 电枢
salient pole 凸极
slip ring 滑环

generator 产生器；发电机
distributed winding 分布式绕组
concentrated winding 集中绕组
squirrel-cage winding 鼠笼绕组
torque 转矩
electromagnetic 电磁的
negative 负的
revolving-field 旋转磁场
axial 轴的
hydroelectric 水力发电的
oscillation 振荡
positive 正的
reactive power 无功功率
mechanical 机械的

Unit 5 Induction Motor

The induction machine is a kind of AC rotational electrical machine, which consists of motors and generators. Three-phase induction motor is widely used in the industry. It is simple, rugged, low priced, durable and robust. At least 90 percent of drives used in the industry are induction motors. They need to run at essentially constant speed from zero to full load. These motors are not easy to be adapted to speed control. However, variable frequency electronic drives are being used more frequently to control the speed of an induction motor.

5.1 Construction of Induction Motors

According to the type of rotor winding, the induction motors are classified into squirrel-cage induction motors and wound-rotor induction motors .

A three-phase induction motor has two main parts: a stationary stator and a revolving rotor. The rotor is separated from the stator by a small air gap that ranges from 0.2 mm to 4 mm, depending on the power of the motor.

5.2 Stator

The stator of an induction machine consists of stator core, stator winding, frame and cover. A steel frame supports a hollow, cylindrical stacked lamination core and provides a space for the stator winding.

5.3 Rotor

The rotor is composed of rotor core, rotor winding and shaft. The rotor core is also composed of punched laminations. These are carefully stacked to create a series of rotor slots in order to provide space for the rotor winding.

A squirrel-cage rotor is composed of bare copper bars, which are slightly longer than the rotor, which are pushed into the slots. The opposite ends are welded into copper end rings, so that all the bars are short-circuited at common points. The entire construction (bars and end rings) resembles a squirrel cage, from which the name is derived. In the same and medium size motors, the bars and the end rings are made of die cast aluminum, which are molded to form an integral block.

A wound rotor consists a complete set of three-phase windings, similar to the one on the stator. The winding is uniformly distributed in the slots and is usually Y-connected. The terminals are connected revolving slip rings and the associated stationary in series with the rotor windings. The external resistors are mainly used during the start-up period. Under a normal running condition, the three brushes are short-circuited.

5.4 Operation Modes of Induction Machines

The operation of a three-phase induction motor is based on the application of Faraday's law and the Lorentz force on a conductor. When the three-phase stator windings are excited from a balanced three-phase source, it will produce a magnetic field in the air gap by rotating at synchronous speed n_s. The synchronous speed is determined by the number of poles and the applied stator frequency f. The revolving field induces a voltage in the rotor bars, and the induced voltage creates large circulating currents which flow in the rotor bars and end rings. The currents carrying rotor bars are immersed in the magnetic field which is created by the stator. They are therefore subjected to a strong mechanical force, and the sum of the mechanical forces on all the rotor bars produces a torque which tends to drag the rotor along the same direction as the revolving field. The rotor speed n is always less than the synchronous speed n_s so as to produce a current in the rotor bars sufficiently large enough to overcome the load torque.

Taking the rotating direction of the revolving magnetic field as a reference, the difference between the synchronous speed n_s and the rotor speed n is named as slip speed Δn.

$$\Delta n = n_s - n \tag{2-19}$$

where Δn is the slip speed; n_s is the speed of the rotating magnetic fields (synchronous speed); n is the rotor speed.

The ratio of the slip speed and the synchronous speed is called slip s (per unit slip).

$$s = \frac{n_s - n}{n_s} \tag{2-20}$$

Slip is an important parameter, which can reveal the operation mode of an induction machine. There are three kinds of operation modes, depending on the rotor speed of induction machines.

5.5 Motor Mode

While the rotor runs at the same direction as a magnetic field and the speed is lower than the synchronous speed, i.e., $0<n<n_s$, then $0<s<1$. The directions of rotor induction electric motive force and the active component of rotor current is shown in Fig. 2-12 (b). The active component of rotor current cooperates with an air-gap magnetic field, the electromagnetic torque is derived as the same direction with the air-gap magnetic field. The rotor and the load rotate at the speed of n. Then the electric energy is converted into the mechanical energy and the induction machine is working as a motor.

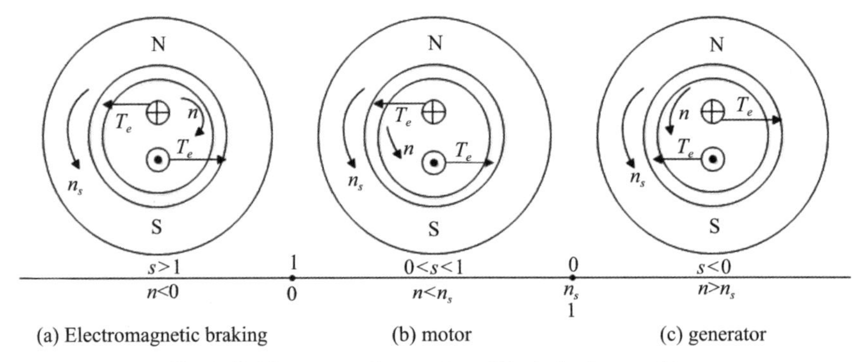

Fig. 2-12 Three operation modes of the induction machine

5.6 Generator Mode

If the rotor speed is higher than the synchronous speed, i.e., $n>n_s$, then $s<0$ (Fig. 2-11 (c)). Comparing to the motor mode, the relative direction of the air-gap magnetic field to the rotor is changed. Then, the directions of the current and the electric motive force of the rotor bar are changed. The direction of the electromagnetic torque is reversed. The induction machine works as a generator and the mechanical energy is converted into the electric energy.

5.7 Electromagnetic Braking

When the direction of the rotor is different from the rotating air-gap field, i.e., $n<0$, then $s>1$ (Fig. 2-12 (a)) , the electromagnetic torque is in the same direction as the air-gap magnetic field, which means that the machine is absorbing electric energy. The direction of the electromagnetic torque is different from the rotor running direction, which means that the machine is absorbing mechanical energy. Both kinds of energy are converted into heat and dissipated into the electrical machine.

Specialized English Words

induction 感应	lamination 叠层
series 串联	resistor 电阻器
synchronous speed 同步转速	electric energy 电能
electromagnetic braking 电磁制动	air gap 气隙

Unit 6 专业英语阅读方法

6.1 专业英语文献阅读方法

专业英语阅读能力是科技工作者所必须具备的基本能力。掌握一定的阅读技巧有助于提高专业英语阅读能力。与普通英文文献的阅读方法相同，专业英语文献常见的阅读方法有四种：Skimming（略读）、Scanning（查读）、Fast reading（快速阅读）、Intensive reading（精读）。其中，前三种属于快速阅读方法。采用哪种阅读方法取决于阅读目的。

Skimming（略读）的目的是快速地获取中心意思和重点。在略读过程中，速度并不总是一致的。当读到与文章主题紧密相关的部分，如段首和段尾的主题句时，可以稍微减慢速度。当读到具体细节部分，如某些例子时，则可以快速掠过。这种阅读方法常用于综述某个研究课题，此时需快速地阅读大量英文文献，总结研究现状。

Scanning（查读）的目的是从文章中迅速查找某一个具体事实或某一项特定信息。在查读过程中，为了快速查找到问题答案，往往不必逐字逐句阅读文章，可以快速地浏览文章，通过找准提示词或定位词发现与问题相关的信息，标记位置，最后再逐一筛选、定位答案。这种阅读方法常用于在大量的资料中快速查找指定信息。

Fast reading（快速阅读）的目的是基本理解文章内容。在速读过程中，要求快速地浏览全文，把握整篇文章的主旨、结构和大概内容，而不必完全看懂文章中每词每句的意思。这种阅读方法常用于细致阅读文章之前的预读、泛读。

Intensive reading（精读）的目的是全面而系统地分析和掌握文章。与普通英语文献的阅读不同，专业英语文献的精读并不是对文章逐词逐句地分析理解，而是对文章中技术细节的精确认识，同时还要了解与文章相关的背景知识。这种方法常用于对从预读中精挑细选出来的文章的精研细读。

在第一章中，我们提到专业英语文献主要包括专业图书、专业论文、科研报告、专利、学位论文及产品说明书等。电气工程与自动化专业的技术人员常常不可避免地要阅读一些英文专业论文和产品说明书。接下来，我们就对英文科技论文和芯片手册的阅读方法逐一进行详细的探讨。

6.2 英文科技论文阅读方法

英文科技论文是作者对其科研领域的最新研究成果或发现的新的研究问题的报告，或者是对某一领域的综述归纳。英文科技论文没有统一的格式。通常，一篇完整的科技论文包括 7 个部分：Title（标题）、Author（作者）、Abstract（摘要）、Keywords（关键词）、Text（正文）、Acknowledgements（致谢）、References（参考文献）。其中，Text（正文）往往采用 IMRaD 形式（Introduction, Materials and Methods, Results, and Discussion）。虽然各个期刊、会议所要求的格式不尽相同，但都有一个固定、规范的行文结构，并且对论文的各个部分的顺序和内容都有严格的要求。正确掌握科技论文的行文结构和各个部分的写作宗旨，有助于读者快速掌握论文内容。

英文科技论文的阅读可采用泛读和精读这两种方法。对于一篇陌生的科技论文，我们可以采用泛读的方式，快速地掌握论文的核心内容。泛读有利于读者筛选适合精读的论文，避免盲目浪费时间阅读不需要了解的文章。

英文科技论文的泛读可以结合运用上述的 Skimming（略读）和 Fast reading（快速阅读）。重点阅读论文的 Abstract（摘要）、Introduction（引言）的最后部分、Results（实验结果）中的 Figure（图）和 Table（表）以及 Discussion（讨论）的结论部分。论文的其余部分可以采用快速阅读的方式大致了解其基本内容。Abstract（摘要）是一篇论文的缩影，有助于读者在短时间内对整篇文章有一个清晰的轮廓认识。Introduction（引言）的最后部分点明论文研究内容和意义。Discussion（讨论）的结论部分强调论文的研究目的和研究内容，并总结研究结果。Results（实验结果）中的 Figure（图）和 Table（表）有助于读者把握论文研究的质量和深度。通过对以上四个部分的重点阅读，读者能在短时间内对整篇文章有一个清晰的轮廓认识，明白论文的哪些部分应该重点理解，哪些部分仅需适当了解，为进一步的精读打下基础。

在对文章有了整体的把握之后，如果觉得该论文有进一步研究价值或对自身的研究有参考意义，就返回采用精读方式通篇进行深入、详细的研究和探讨。最后是阅读参考文献，顺藤摸瓜，找出该领域的重点文献，以便之后系统、集中地阅读相关文献。

6.3 英文芯片手册的阅读方法

芯片手册，即芯片的使用说明书，其最显著的一个特点就是必须尽可能地使用通俗易懂的语句，向使用者交代清楚该产品的特点、功能以及使用方法，而较少使用偏僻的语法或专业词汇以外的生僻单词（当然专业词汇除外）。一般来说，运用在大学里所学到的英文知识足以阅读分析芯片手册。

此外，为保证严谨，不至于让读者产生误解，数据手册通常喜欢采用一些长句对问题进行描述，并且这些长句所描述的问题通常都比较关键。理解这些长句可以采用结构分析法，按照主谓宾状补结构，把整个长句拆开，然后对每一个小短句进行分析，最后联系上下文，揣摩出整句的意思。而且，芯片手册往往使用较多的专业词汇，有些专业词汇甚至在专业字典里都无从查找。这时，我们可以借助网络资源，比如 CNKI 翻译助手等。需要强调的是，相较其他的专业英语文献，如期刊论文，芯片手册的上下文逻辑性与连贯性不强，句与句之间的过渡较为生硬。因此，不必把每一个单词的意思都完完全全、准确无误地翻译出来，也无须执着理清句与句之间的逻辑关系，只要理解它所表达的意思，能够正常使用芯片手册即可。

一般来说，阅读芯片的数据手册可采用如下步骤：

首先，阅读芯片的特性（Features）、应用场合（Applications）以及内部框图，以便对芯片有一个宏观的了解。同时，标注该芯片的特殊功能，充分利用芯片的特殊功能将会对电路的整体设计有极大的好处。比如 AD9945 可以实现相关双采样（CDS），这可以简化后续信号调理电路，加强抵抗噪声的效果。

其次，重点关注芯片的参数，同时可以参考手册给出的一些参数图（如 AD9945 的 TPC1、TPC2 等），这是是否采用该芯片的重要依据。例如，对于 AD9945，就可以关注采样率（maximum clock rate）、数据位数（AD converter）、功耗（power consumption）、可调增益范围（gain range）等。

选定器件后，研究芯片管脚的定义、手册推荐的 PCB layout 等，这些都是在硬件设计过程中必须掌握的。所有管脚中，要特别留意控制信号引脚或者特殊信号引脚，这是将来用好该芯片的前提。比如 AD9945 的 SHP、SHD、PBLK、CLPOB 等。

认真研读芯片内部寄存器。对寄存器的理解程度将直接决定对该芯片的掌握程度。比如 AD9945 有 4 个寄存器：Operation、Control、Clamp Level 和 VGA gain。对于这些寄存器，必须清楚它们上电后的初始值、所能实现的功能、每个 bit 所代表的含义等基本情况。

最后，仔细研究手册给出的时序图，这是对芯片进行正确操作的关键。单个信号的周期、上升时间、下降时间、建立时间、保持时间，以及信号之间的相位关系，所有这些都必须研究透彻。像 AD9945 的 Figure 8 和 Figure 9 就很值得花费时间去仔细研究。

需要注意的是，凡是芯片数据手册中的“note”，都必须仔细阅读，这也是能正确使用或把芯片用好的关键所在。

实训课程二：专业英语阅读

1. 课文阅读练习

1-1 What is Ampere's law?

1-2 What is magnetizing intensity? What is magnetic flux density? How are they related?

1-3 What is Faraday's law?

1-4 What conditions are necessary for a magnetic field to produce a force on a wire?

1-5 What conditions are necessary for a magnetic field to produce a voltage in a wire?

2-1 What is commutation? How can a commutator convert AC voltages on a machine's armature to DC voltages at its terminals?

2-2 What is the difference between a DC machine and an AC machine?

2-3 How many windings are there in a DC machine? What are they, and what is their role?

2-4 What does the DC motor consist of?

2-5 Where does the driving torque of the coil in the generator come from?

3-1 Is the turns ratio of a transformer the same as the ratio of voltages across the transformer ? Why or why not?

3-2 Describe the construction of transformers.

3-3 List the purposes of transformers.

3-4 What is the leakage flux in a transformer ? Why is it modeled in a transformer equivalent circuit as an inductor?

3-5 List and describe the types of losses that occur in a transformer. Why does the power factor of a load affect the voltage regulation of a transformer?

4-1 Describe the difference between a stationary-field synchronous machine and a revolving-field synchronous machine.

4-2 How many types are the revolving-field synchronous machines classified into? What are they?

4-3 Why are the field windings also called rotor windings?

4-4 Explain when the damper winding does not carry any current.

4-5 List and explain the important variables which reflect operation modes of an induction machine.

5-1 What are the two parts of the induction motor?

5-2 How does the induction motor generate a rotating magnetic field?

5-3 What are the working states of induction motors?

5-4 How to change the speed of the synchronous machine?

5-5 Why does the induction motor need a balanced 3-phase source?

5-6 What are the slip and the slip speed in an induction motor?

5-7 What is the connection of the windings of a wound rotor?

2. LM555 芯片阅读练习

LM555 是一个定时器芯片，外接电阻和电容后既可以作为单稳态触发器，也可以作为多谐振荡器。常用在比较器、触发器、电压分配器、功率输出级等功能模块中。可以通过 http://www.alldatasheetcn.com、http://www.ic37.com 或芯片的官网 www.national.com 查找芯片 LM555 的 Datasheet（数据表）。

（1）芯片的特性：芯片的正常和异常的电气特性。

Features

- Direct replacement for SE555/NE555
- Timing from microseconds through hours
- Operates in both astable and monostable modes
- Adjustable duty cycle
- Output can source or sink 200 mA
- Output and supply TTL compatible
- Temperature stability better than 0.005% per °C
- Normally on and normally off output
- Available in 8-pin MSOP package

（2）芯片的应用场合：器件的应用领域，通常概括性地罗列内容。

Applications

- Precision timing
- Pulse generation
- Sequential timing
- Time delay generation
- Pulse width modulation
- Pulse position modulation
- Linear ramp generator

（3）芯片内部框图：通常被称作等效原理图。

Schematic Diagram

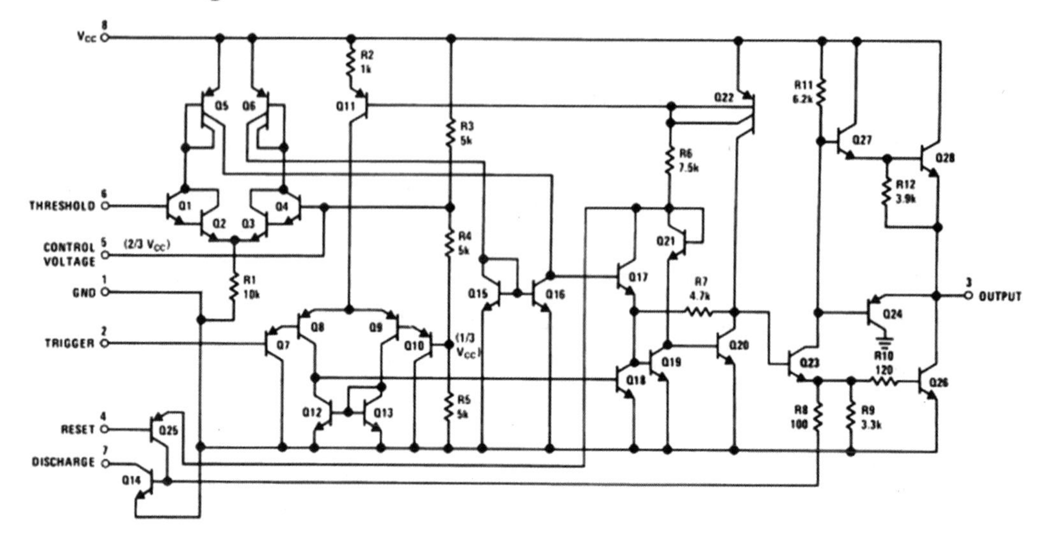

Chapter 3
POWER ELECTRONICS

Unit 1 Power Electronic Devices

1.1 Power Semiconductor Devices

Since the first thyristor SCR was developed in late 1957, there have been tremendous advances in the power semiconductor devices. Until 1970, the conventional thyristors had been exclusively used for power control in industrial applications. Since 1970, various types of power semiconductor devices have been developed and become commercially available. Fig. 3-1 (omitted) shows the classification of the power semiconductors, which are made of either silicon or silicon carbide. Silicon carbide devices are, however, under development. A majority of the devices are made of silicon. These devices can be further divided into three types: (1) power diodes, (2) transistors, and (3) thyristors. These can be further divided broadly into five types: (1) power diodes, (2) thyristors, (3) power bipolar junction transistors (BJTs), (4) power metal oxide semiconductor field effect transistors (MOSFETs), and (5) insulated gate bipolar transistors (IGBTs) and static induction transistors (SITs).

1.2 Power Diodes

A diode has two terminals: a cathode and an anode. Power diodes are of three types: general purpose, high speed (or fast recovery), and Schottky. General purpose diodes are available up to 6000 V, 4500 A, and the rating of fast recovery diodes can go up to 6000 V, 1100 A. The reverse recovery time varies between 0.1 and 5 us. Fast recovery diodes are essential for high frequency switching of power converters. Schottky diodes have low on-state voltage and very short recovery time, typically nanoseconds. The leakage current increases with the voltage rating and their ratings are limited to 100 V, 300 A. A diode conducts when its anode voltage is higher than that of the cathode; and the forward voltage drop of a power diode is very low, typically between 0.5 and 1.2 V. If the cathode voltage is higher than its anode voltage, a diode is said to be in a blocking.

1.3 Thyristors

A thyristor has three terminals: an anode, a cathode, and a gate. When a small current is passed through the gate terminal to the cathode, the thyristor conducts, provided that the anode terminal is at a higher potential than the cathode. The thyristors can be subdivided into eleven types: (a) forced commutated thyristor, (b) line commutated thyristor, (c) gate turn off thyristor (GTO), (d) reverse conducting thyristor (RCT), (e) static induction thyristor (SITH), (f) gate assisted turn-off thyristor (GATT), (g) light activated silicon controlled rectifier (LASCR), (h) MOS turn-off (MTO) thyristor, (i) emitter turn-off (ETO) thyristor, (j) integrated gate-commutated thyristor (IGCT), and (k) MOS controlled thyristors (MCTs). Once a thyristor is in a conduction mode, the gate circuit has no control and the thyristor continues to conduct. When a thyristor is in a conduction mode, the forward voltage drop is very small, typically 0.5 to 2 V. A conducting thyristor can be turned off by making the potential of the anode equal to or less than that of the cathode. The line commutated thyristors are turned off due to the sinusoidal nature of the input voltage, and forced commutated thyristors are turned off by an extra circuit called commutation circuitry.

Natural or line commutated thyristors are available with ratings up to 6000 V, 4500 A. The *turn-off time* of high-speed reverse-blocking thyristors has been improved substantially and it is possible to have 10 to 20 us in a 3000 V, 3600 A thyristor. The turn-off time is defined as the time interval between the instant when the principal current has decreased to zero after external switching of the principal voltage circuit, and the instant when the thyristor is capable of supporting a specified principal voltage without turning on. RCTs and GATTs are widely used for high-speed switching, especially in traction applications. An RCT can be considered as a thyristor with an inverse parallel diode. RCTs are available up to 4000 V, 2000 A (and 800 A in reverse conduction) with a switching time of 40 us. GATTs are available up to 1200 V, 400 A with a switching speed of 8 us. LASCRs, which are available up to 6000 V, 1500 A with a switching speed of 200 to 400 us, are suitable for high voltage power systems, especially in HVDC. For low power AC applications, TRIACs are widely used in all types of simple heat controls, light controls, motor controls, and AC switches. The characteristics of TRIACs are similar to two thyristors connected in inverse parallel and having only one gate terminal. The current flow through a TRIAC can be controlled in either direction.

GTOs and SITHs are self-turned-off thyristors. GTOs and SITHs are turned on by applying a short positive pulse to the gates and are turned off by the applications of a short negative pulse to the gates. They do not require any commutation circuit. GTOs are very attractive for forced commutation of converters and are available up to 6000 V, 6000 A. SITHs, whose ratings can go as high as 1200 V, 300 A, are expected to be applied for medium-power converters with a frequency of several hundred kilohertz and beyond the frequency range of GTOs. Fig. 3-2 (omitted) shows various configurations of GTOs. An MTO is a combination of a GTO and a MOSFET, which together overcome the limitations of the GTO turn-off ability. Its structure is similar to that of a GTO and retains the GTO advantages of high voltage (up to 10 kV) and high current (up to 4000 A). MTOs can be used in high power applications ranging from 1 to 20 MV · A. An ETO is a MOS-GTO hybrid device that combines the advantages of both the GTO and the MOSFET. ETO has two gates: one normal gate for turn-on and one with a series MOSFET for turn-off. ETOs with a current rating of up to 4 kA and a voltage rating of up to 6 kV have been demonstrated.

An IGCT integrates a gate commutated thyristor (GCT) with a multilayered gate driver circuit board. The GCT is a hard switched GTO with a very fast and large gate current pulse, as large as the full rated current that draws out all the current from the cathode into the gate in about 1 us to ensure a fast turn-off. Similar to a GTO, the IGCT is turned on by applying the turn-on current to its gate. The IGCT is turned off by a multilayered gate driver circuit board that can supply a fast-rising turn-off pulse (i.e., a gate current of 4 kA/us with gate cathode voltage of 20 V only). A MCT can be turned "on" by a small negative voltage pulse on the MOS gate (with respect to its anode), and turned "off" by a small positive voltage pulse. It is like a GTO, except that the turn-off gain is very high. MCTs are available up to 4500 V, 250 A.

1.4 Power Transistors

Power transistors are of four types: (1) BJTs, (2) power MOSFETs, (3) IGBTs, and (4) SITs. A bipolar transistor has three terminals: base, emitter, and collector. It is normally operated as a switch in the common emitter configuration. As long as the base of an NPN transistor is at a higher potential than the emitter and the base current is sufficiently large to drive the transistor in the saturation region, the transistor remains on, provided that the collector-to-emitter junction is properly biased. High power bipolar transistors are commonly used in power converters at a frequency below 10 kHz and are effectively applied in the power ratings up to 1200 V, 400 A. The forward drop of a conducting transistor is in the range 0.5 to 1.5 V. If the base drive voltage is withdrawn, the transistor remains in the nonconduction (or off) mode.

Power MOSFETs are used in high speed power converters and are available at a relatively low power rating in the range of 1000 V, 100 A at a frequency range of several tens of kHz. IGBTs are voltage-controlled power transistors. They are inherently faster than BJTs, but still not quite as fast as MOSFETs. However, they offer far superior drive and output characteristics to those of BJTs. IGBTs are suitable for high voltage, high current, and frequencies up to 20 kHz. IGBTs are available up to 1700 V, 2400 A.

COOLMOS is a new technology for high voltage power MOSFETs, and it implements a compensation structure in the vertical drift region of a MOSFET to improve the on-state resistance. It has a lower on-state resistance for the same package compared to that of other MOSFETs. The conduction losses are at least 5 times less as compared to those of the conventional MOSFET technology. COOLMOS is capable of handling two to three times more output power as compared to the conventional MOSFET in the same package. The active chip area of COOLMOS is approximately 5 times smaller than that of a standard MOSFET. The on-state resistance of a 600 V, 47 A COOLMOS is 70 mΩ.

A SIT is a high power, high frequency device. It is essentially the solid state version of the triode vacuum tube, and is similar to a junction field effect transistor (JFET). It has a low noise, low distortion, high audio frequency power capability. The turn-on and turn-off times are very short, typically 0.25 us. The normally on-characteristic and the high on-state drop limit its applications for general power conversions. The current rating of SITs can be up to 1200 V, 300 A, and the switching speed can be as high as 100 kHz. SITs are most suitable for high power, high frequency applications (e.g., audio, VHF/ultrahigh frequency [UHF], and microwave amplifiers).

Specialized English Words

thyristor 晶闸管
semiconductor 半导体
transistor 三极管；晶体管
anode 正极；阳极
converter 转换器
leakage current 漏电流
commutate 变换电流方向
traction 牵引；牵引力
tremendous 极大的；巨大的
diode 二极管
cathode 阴极，负极
Schottky 肖特基
on-state 开态；通路状态
subdivide 细分；再分
sinusoidal 正弦的
saturation 饱和
bipolar junction transistor（BJT） 双极型晶体管
field effect transistor（FET） 场效应晶体管
insulated gate bipolar transistor（IGBT） 绝缘栅双极晶体管
silicon controlled rectifier（SCR） 可控硅整流器

Unit 2 Rectifier

2.1 Introduction

Diodes are extensively used in rectifiers. A *rectifier* is a circuit that converts an AC signal into a unidirectional signal. A rectifier is a type of AC-DC converter. Depending on the type of input supply, the rectifiers are classified into two types: (1) single phase and (2) three phase. For the sake of simplicity the diodes are considered to be ideal. By "ideal" we mean that the reverse recovery time and the forward voltage drop are negligible. That is, $t_{rr} = 0$ and $V_D = 0$.

2.2 Single-Phase Half-Wave Rectifiers

A single-phase half-wave rectifier is the simplest type, but it is not normally used in industrial applications. However, it is useful in understanding the principle of rectifier operation. The circuit diagram with a resistive load is shown in Fig. 3-3 (a) (omitted). During the positive half-cycle of the input voltage, diode D_1 conducts and the input voltage appears across the load. During the negative half-cycle of the input voltage, the diode is in a *blocking condition* and the output voltage is zero. The waveforms for the input voltage and output voltage are shown in Fig. 3-3 (b) (omitted).

2.3 Performance Parameters

Although the output voltage as shown in Fig. 3-3 (b) (omitted) is DC, it is discontinuous and contains harmonics. A rectifier is a power processor that should give a DC output voltage with a minimum amount of harmonic contents. At the same time, it should keep the input current as sinusoidal as possible and in phase with the input voltage so that the power factor is near unity. The power processing quality of a rectifier requires the determination of harmonic contents of the input current, the output voltage, and the output current. We can use Fourier series expansions to find the harmonic contents of voltages and currents. There are different types of rectifier circuits and the performances of a rectifier are normally evaluated in terms of the following parameters:

(1) The average value of the output (load) voltage, V_{dc};

(2) The average value of the output (load) current, I_{dc} ;

(3) The output DC power, $P_{dc}=V_{dc}I_{dc}$;

(4) The root mean square (rms) value of the output voltage, V_{rms} ;

(5) The rms value of the output current, I_{rms} ;

(6) The output AC power, $P_{ac}=V_{rms}I_{rms}$.

2.4 Single-Phase Full-Wave Rectifiers

A full-wave rectifier circuit with a center tapped transformer is shown in Fig. 3-4 (a) (omitted). Each half of the transformer with its associated diode acts as a half-wave rectifier and the output of a full-wave rectifier is shown in Fig. 3-3 (b) (omitted). Because there is no DC current flowing through the transformer, there is no DC saturation problem of transformer core. The average output voltage is

$$V_{dc}=\frac{2}{T}\int_0^{T/2} V_m \sin \omega t dt = \frac{2V_m}{\pi} = 0.6366V_m \tag{3-1}$$

Instead of using a center tapped transformer, we could use four diodes, as shown in Fig. 3-5 (a) (omitted). During the positive half-cycle of the input voltage, the power is supplied to the load through diodes D_1 and D_2. During the negative cycle, diodes D_3 and D_4 conduct. The waveform for the output voltage is shown in Fig. 3-6 (omitted) and is similar to that of Fig. 3-5 (b) (omitted). The peak inverse voltage of a diode is only V_m. This circuit is known as a *bridge rectifier,* and it is commonly used in industrial applications.

Specialized English Words

rectifier 整流器

voltage drop 电压降

positive 正的

harmonic 谐波

transformer 变压器

direct current（DC） 直流

forward 正向的

waveform 波形

negative 负的

power factor 功率因数

saturation 饱和

alternating current（AC） 交流

Unit 3 Inverter

3.1 Introduction

DC-to-AC converters are known as *inverters*. The function of an inverter is to change a DC input voltage to a symmetric AC output voltage of desired magnitude and frequency. The output voltage could be fixed or variable at a fixed or variable frequency. A variable output voltage can be obtained by varying the input DC voltage and maintaining the gain of the inverter constant. On the other hand, if the DC input voltage is fixed and it is not controllable, a variable output voltage can be obtained by varying the gain of the inverter, which is normally accomplished by pulse width modulation (PWM) control within the inverter. The *inverter gain* may be defined as the ratio of the AC output voltage to DC input voltage.

The output voltage waveforms of ideal inverters should be sinusoidal. However, the waveforms of practical inverters are nonsinusoidal and contain certain harmonics. For low and medium power applications, square wave or quasi-square wave voltages may be acceptable; and for high power applications, low distortion sinusoidal waveforms are required. With the availability of high speed power semiconductor devices, the harmonic contents of output voltage can be minimized or reduced significantly by switching techniques.

Inverters are widely used in industrial applications (e.g, variable speed AC motor drives, induction heating, standby power supplies, and uninterruptible power supplies). The input may be a battery, fuel cell, solar cell, or some other DC source. The typical single-phase outputs are (1) 120 V at 60 Hz, (2) 220 V at 50 Hz, and (3) 115 V at 400 Hz. For high power three-phase systems, typical outputs are (1) 220 to 380 V at 50 Hz, (2) 120 to 208 V at 60 Hz, and (3) 115 to 200 V at 400 Hz.

Inverters can be broadly classified into two types: (1) single-phase inverters, and (2) three-phase inverters. Each type can use controlled turn-on and turn-off devices. These inverters generally use PWM control signals for producing an AC output voltage. An inverter is called a *voltage-fed inverter* (VFI) if the input voltage remains constant, a *current-fed inverter* (CFI) if the input current is maintained constant, and a *variable DC linked inverter* if the input voltage is controllable. If the output voltage or current of the inverter is forced to pass through zero by creating an LC resonant circuit, this type of inverter is called the *resonant-pulse inverter* and it has wide applications in power electronics.

3.2 Principle of Operation

The principle of single-phase inverters can be explained with Fig. 3-7 (omitted). The inverter circuit consists of two choppers. When only transistor Q_1 is turned on for a time $T_0/2$, the instantaneous voltage across the load v_0 is $V_S/2$. If only transistor Q_2 is turned on for a time $T_0/2$, $-V_S/2$ appears across the load. The logic circuit should be designed such that Q_1 and Q_2 are not turned on at the same time. This inverter requires a three-wire DC source, and when a transistor is off, its reverse voltage is V_S instead of $V_S/2$. This inverter is known as a *half-bridge inverter.*

The root mean square (rms) output voltage can be found from

$$V_0 = \left(\frac{2}{T} \int_0^{T_0/2} \frac{V_s^2}{4} \mathrm{d}t \right)^{1/2} = \frac{V_s}{2} \tag{3-2}$$

The instantaneous output voltage can be expressed in Fourier series as

$$v_0 = \frac{a_0}{2} + \sum_{n=1}^{\infty} \left(a_n \cos(n\omega t) + b_n \sin(n\omega t) \right) \tag{3-3}$$

Due to the quarter-wave symmetry along the x-axis, both a_0 and a_n are zero. We get b_n as

$$b_n = \frac{1}{\pi} \left(\int_{\frac{-\pi}{2}}^{0} \frac{-V_s}{2} \mathrm{d}(\omega t) + \int_0^{\frac{\pi}{2}} \frac{V_s}{2} \mathrm{d}(\omega t) \right) = \frac{4V_s}{n\pi} \tag{3-4}$$

which gives the instantaneous output voltage v_0 as

$$v_0 = \sum_{n=1,3,5,\ldots}^{\infty} \frac{2V_s}{n\pi} \sin n\omega t = 0 \qquad \text{for } n = 2,4,\ldots \tag{3-5}$$

Where $\omega = 2\pi f_0$ is the frequency of output voltage in rads per second. Due to the quarter-wave symmetry of the output voltage along the x-axis, the even harmonics voltages are absent. For $n = 1$, Eq. (3-6) gives the rms value of fundamental component as

$$v_{01} = \frac{2V_s}{\sqrt{2}\pi} = 0.45V_s \tag{3-6}$$

For an inductive load, the load current cannot change immediately with the output voltage. If Q_1 is turned off at $t = T_0/2$, the load current would continue to flow through D_2 load, and the lower half of the DC source until the current falls to zero. Similarly, when Q_1 is turned off at $t = T_0$, the load current flows through D_1 load, and the upper half of the DC source. When diode D_1 or D_2 conducts, energy is fed back to the DC source and these diodes are known as *feedback diodes.* Fig. 3-7 (c) (omitted) shows the load current and conduction intervals of devices for a purely inductive load. It can be noticed that for a purely inductive load, a transistor conducts only for $T_0/2$ (or 90°). Depending on the load impedance angle, the conduction period of a transistor would vary from 90° to 180° .

Any switching devices can replace the transistors. If t_{off} is the turn-off time of a device, there must be a minimum delay time of $t_d (= t_{on})$ between the outgoing device and triggering of the next incoming device. Otherwise, short-circuit condition would result through the two devices. Therefore, the maximum conduction time of a device would be $t_{on} = T_0/2 - t_d$. All practical devices require a certain turn-on and turn-off time. For successful operation of inverters, the logic circuit should take these into account.

For an RL load, the instantaneous load current i_0 can be found by dividing the instantaneous output voltage by the load impedance $Z = R + jn\omega L$. Thus, we get

$$i_0 = \sum_{n=1,3,5,\ldots}^{\infty} \frac{2V_s}{n\pi\sqrt{R^2 + (n\omega L)^2}} \sin(n\omega t - \theta_n) \tag{3-7}$$

where $\theta_n = \tan^{-1}(n\omega L/R)$. If I_{01} is the rms fundamental load current, the fundamental output power (for $n = 1$) is

$$\begin{aligned} P_{01} &= V_{01} I_{01} \cos\theta_1 = I_{01}^2 R \\ &= \left(\frac{2V_s}{\sqrt{2}\pi\sqrt{R^2 + (\omega L)^2}} \right)^2 R \end{aligned} \tag{3-8}$$

3.3 Performance Parameters

The output of practical inverters contain harmonics and the quality of an inverter is normally evaluated in terms of the following performance parameters.

Harmonic factor of nth harmonic (HF_n). The harmonic factor (of the nth harmonic), which is a measure of individual harmonic contribution, is defined as

$$HF_n = \frac{V_{on}}{V_{ol}} \quad \text{for } n > 1 \tag{3-9}$$

where V_{ol} is the rms value of the fundamental component and V_{on} is the rms value of the nth harmonic component.

Total harmonic distortion (THD). The total harmonic distortion, which is a measure of closeness in shape between a waveform and its fundamental component, is defined as

$$THD = \frac{1}{V_{ol}} \left(\sum_{n=2,3,\ldots}^{\infty} V_{on}^2 \right)^{1/2} \tag{3-10}$$

Distortion factor (DF). THD gives the total harmonic content, but it does not indicate the level of each harmonic component. If a filter is used at the output of inverters, the higher order harmonics would be attenuated more effectively. Therefore, a knowledge of both the frequency and magnitude of each harmonic is important. The DF indicates the amount of THD that remains in a particular waveform after the harmonics of that waveform have been subjected to a second order attenuation (i.e., divided by n^2). Thus, DF is a measure of effectiveness in reducing unwanted harmonics without having to specify the values of a second order load filter and is defined as

$$DF = \frac{1}{V_{\text{ol}}}\left(\sum_{n=2,3,\ldots}^{\infty}\left(\frac{V_{\text{on}}}{n^2}\right)^2\right)^{1/2} \tag{3-11}$$

The DF of an individual (or nth) harmonic component is defined as

$$DF_n = \frac{V_{\text{on}}}{V_{\text{ol}}n^2} \quad \text{for } n>1 \tag{3-12}$$

Lowest order harmonic (LOH). The LOH is the harmonic component whose frequency is closest to the fundamental one, and its amplitude is greater than or equal to 3% of the fundamental component.

Specialized English Words

inverter 反相器；变流器
gain 增益
chopper 斩波器
pulse width modulation (PWM) 脉冲宽度调制
uninterruptible power supply (UPS) 不间断电源
symmetric 对称的；匀称的
distort 扭曲；变形
delay time 滞后时间；延迟时间

Unit 4 DC Converter

4.1 Introduction

In many industrial applications, it is required to convert a fixed voltage DC source into a variable voltage DC source. A DC-DC converter converts directly from DC to DC and is simply known as a DC converter. A DC converter can be considered as DC equivalent to an AC transformer with a continuously variable turns ratio. Like a transformer, it can be used to step down or step up a DC voltage source.

DC converters are widely used for traction motor control in electric automobiles, trolley cars, marine hoists, forklift trucks, and mine haulers. They provide smooth acceleration control, high efficiency, and fast dynamic response. DC converters can be used in regenerative braking of DC motors to return energy back into the supply, and this feature results in energy savings for transportation systems with frequent stops. DC converters are used in DC voltage regulators; and also are used, in conjunction with an inductor, to generate a DC current source, especially for the current source inverter.

4.2 Principle of Step-Down Operation

The principle of operation can be explained by Fig. 3-8 (a) (omitted). When switch SW, known as the chopper, is closed for time t_1 the input voltage appears across the load. If the switch remains off for time t_2 the voltage across the load is zero. The waveforms for the output voltage and load current are also shown in Fig. 3-8 (b) (omitted). The converter switch can be implemented by using (1) a power bipolar junction transistor (BJT), (2) a power metal oxide semiconductor field-effect transistor (MOSFET), (3) a gate-turn-off thyristor (GTO), or (4) an insulated-gate bipolar transistor (IGBT).

The practical devices have a finite voltage drop ranging from 0.5 to 2 V, and for the sake of simplicity we shall neglect the voltage drops of these power semiconductor devices.

The average output voltage is given by

$$V_a = \frac{1}{T}\int_0^{t_1} v_0 \mathrm{d}t = \frac{t_1}{T}V_s = ft_1V_s = kV_s \tag{3-13}$$

and the average load current, $I_a=V_a/R=kV_s/R$, where T is the chopping period; $k = t_1/T$ is the duty cycle of chopper; f is the chopping frequency.

The rms value of output voltage is found from

$$V_0 = \left(\frac{1}{T}\int_0^{kT} {v_0}^2 \mathrm{d}t\right)^{1/2} = \sqrt{k}V_s \tag{3-14}$$

Assuming a lossless converter, the input power to the converter is the same as the output power and is given by

$$P_i = \frac{1}{T}\int_0^{kT} v_0 i \mathrm{d}t = \frac{1}{T}\int_0^{kT} \frac{v_0^2}{R} \mathrm{d}t = k\frac{V_s^2}{R} \tag{3-15}$$

The effective input resistance seen by the source is

$$R_i = \frac{V_s}{I_a} = \frac{V_s}{kV_s/R} = \frac{R}{k} \tag{3-16}$$

which indicates that the converter makes the input resistance R_i as a variable resistance of R/k.

The duty cycle k can be varied from 0 to 1 by varying T, or f. Therefore, the output voltage V_0 can be varied from 0 to V_s by controlling k, and the power flow can be controlled.

(1) *Constant-frequency operation:* The converter, or switching, frequency f (or chopping

period T) is kept constant and the on-time t_1 is varied. The width of the pulse is varied and this type of control is known as *pulse width modulation* (PWM) control.

(2) *Variable-frequency operation:* The chopping, or switching, frequency f is varied. Either on-time t_1 or off-time t_2 is kept constant. This is called *frequency modulation.* The frequency has to be varied over a wide range to obtain the full output voltage range. This type of control would generate harmonics at unpredictable frequencies and the filter design would be difficult.

4.3 Principle of Step-Up Operation

A converter can be used to step up a DC voltage and an arrangement for step-up operation is shown in Fig. 3-10 (a) (omitted). When switch SW is closed for time t_1 the inductor current rises and energy is stored in the inductor L. If the switch is opened for time t_2 the energy stored in the inductor is transferred to load through diode D_1 and the inductor current falls. Assuming a continuous current flow, the waveform for the inductor current is shown in Fig. 3-10 (b) (omitted).

When the converter is turned on, the voltage across the inductor is

$$V_{\mathrm{L}} = L\frac{\mathrm{d}i}{\mathrm{d}t} \tag{3-17}$$

and this gives the peak-to-peak ripple current in the inductor as

$$\Delta I = \frac{V_s}{L}t_1 \tag{3-18}$$

The average output voltage is

$$V_0 = V_s + L\frac{\Delta I}{t_2} = V_s(1+\frac{t_1}{t_2}) = V_s\frac{1}{1-k} \tag{3-19}$$

If a large capacitor C_{L} is connected across the load as shown by dashed lines in Fig. 3-10 (a) (omitted), the output voltage is continuous and v_0 becomes the average value V_a. We can notice from Eq. (3-19) that the voltage across the load can be stepped up by varying the duty cycle k and the minimum output voltage is V_s when k=0. However, the converter cannot be switched on continuously such that k=1. For values of k: tending to unity, the output voltage becomes very large and is very sensitive to changes in k, as shown in Fig. 3-10 (c) (omitted).

4.4 Performance Parameters

The power semiconductor devices require a minimum time to turn on and turn off. Therefore, the duty cycle k can only be controlled between a minimum value $k_{\min}$ and a maximum value $k_{\max}$, thereby limiting the minimum and maximum value of output voltage. The switching frequency of the converter is also limited. It can be noticed that the load ripple current depends inversely on the chopping frequency f. The frequency should be as high as possible to reduce the load ripple current and to minimize the size of any additional series inductor in the load circuit.

The performance parameters of the step-up and step-down converters are as follows:

(1) Ripple current of the inductor, ΔI_L;

(2) Maximum switching frequency, f_{max};

(3) Condition for continuous or discontinuous inductor current;

(4) Minimum value of inductor to maintain continuous inductor current;

(5) Total harmonic distortion, THD.

Specialized English Words

dynamic　动态的

regenerative braking　再生制动

regulator　调节器

inductor　电感器

capacitor　电容器

duty cycle　占空比

switching frequency　开关频率

ripple current　纹波电流

total harmonic distortion (THD)　总谐波失真

Unit 5　AC Voltage Controllers

5.1 Introduction

If a thyristor switch is connected between AC supply and load, the power flow can be controlled by varying the RMS value of AC voltage applied to the load; and this type of power circuit is known as an AC voltage controller. The most common applications of AC voltage controllers are: industrial heating, on-load transformer connection changing, light controls, speed control of polyphase induction motors, and AC magnet controls. For power transfer, two types of control are normally used: (1) On-off control; and (2) Phase angle control.

In on-off control, thyristor switches connect the load to the AC source for a few cycles of input voltage and then disconnect it for another few cycles. In phase control, thyristor switches connect the load to the AC source for a portion of each cycle of input voltage.

The AC voltage controllers can be classified into two types: (1) single phase controllers and (2) three phase controllers, with each type subdivided into (a) unidirectional or half-wave control and (b) bidirectional or full-wave control. There are various configurations of three-phase controllers depending on the connections of thyristor switches.

Because the input voltage is AC, thyristors are line commutated; and phase control thyristors, which are relatively inexpensive and slower than fast switching thyristors, are normally used. For applications up to 400 Hz, if TRIACs are available to meet the voltage and current ratings of a particular application, TRIACs are commonly used.

Due to line or natural commutation, there is no need of extra commutation circuitry and the circuits for AC voltage controllers are very simple. Due to the nature of output waveforms, the analysis for the derivations of explicit expressions for the performance parameters of circuits is not simple, especially for phase-angle-controlled converters with RL loads. For the sake of simplicity, resistive loads are considered in this chapter to compare the performances of various configurations. However, the practical loads are of the RL type and should be considered in the design and analysis of AC voltage controllers.

5.2 Principle of On-Off Control

The principle of on-off control can be explained with a single-phase full-wave controller. The thyristor switch connects the AC supply to load for time t_n the switch is turned off by a gate pulse inhibiting for time t_0. The on-time t_n usually consists of an integral number of cycles. The thyristors are turned on at the zero voltage crossings of AC input voltage. The gate pulses for thyristors and T_2 and the waveforms for input and output voltages are shown in Fig. 3-11 (omitted).

This type of control is applied in applications that have a high mechanical inertia and high thermal time constant (e.g., industrial heating and speed control of motors). Due to zero voltage and zero current switching of thyristors, the harmonics generated by switching actions are reduced.

5.3 Principle of Phase Control

The principle of phase control can be explained with reference to Fig. 3-12 (a) (omitted). The power flow to the load is controlled by delaying the firing angle of thyristor T_1. Fig. 3-12 (b) (omitted) illustrates the gate pulses of thyristor T_1 and the waveforms for the input and output voltages. Due to the presence of diode D_1 the control range is limited and the effective RMS output voltage can only be varied between 70.7% and 100%. The output voltage and input current are asymmetric and contain a DC component. If there is an input transformer, it may cause a saturation problem. This circuit is a single-phase half-wave controller and is suitable only for low power resistive loads, such as heating and lighting. Because the power flow is controlled during the positive half-cycle of input voltage, this type of controller is also known as a *unidirectional controller*.

Specialized English Words

root mean square (rms) 均方根

full wave 全波

unidirectional 单方向的

controller 控制器

Unit 6 科技英语口译

6.1 科技口译的特点

科技口译是当今国内科技市场与国际科技市场衔接急需的一种专业技能。它要求科技人员既能读懂英文文献，又能将英文文献中的信息实时传递给其他与会人员。相较于普通英语口译，科技口译的特点可归纳如下：

1. 时效性

常见翻译，例如笔译，分为理解、翻译、表达三个阶段。然而，科技口译的特殊性在于瞬时性，这三个阶段都要在短时间完成，没有时间查阅书籍和其他材料。

2. 专业性

科技口译涉及的都是专业问题，专业英语翻译要求准确、精炼，因此，要求口译人员熟悉该领域的知识，并掌握该领域的常用专业词汇。特别是，在口译过程中常遇到不少缩略语，口译人员既要注重平时搜集和整理缩略语，也要不断掌握新出现的缩略语。

3. 独立性

在口译过程中，口译人员必须完全独立处理遇到的问题。为了使信息接收人员能够接收到准确无误的翻译信息，口译人员必须掌握较全面的专业知识和翻译术语，并且具备一些口译技巧和实战经验。

6.2 科技英语口译方法

1. 掌握专业英语特点

如上所述，科技口译一般都要在短时间内完成，熟练掌握专业英语和普通英语在语言表达上的区别，有助于加快完成科技口译的速度。例如，科技英文文献常常使用被动语态，但在汉语中使用被动语态的频率要远低于英语。因此，进行科技口译时，将英语的被动语态转换为汉语的主动语态是常见的翻译手法。此外，还有诸如时间状语的位置、因果逻辑等在英语和汉语中的表达很不一样。明确这些区别，我们就可以方便地完成科技口译。

2. 加强理解分析

理解是翻译的第一个阶段。逻辑在科技英语中占有重要的位置。因此，快速理清科技英语中的逻辑关系是短时间内准确翻译科技英语的关键。科技英语中的逻辑关系分为纵向关系（即文章整体架构关系）和横向关系，即句与句之间的因果、转折、并列、主次等逻辑关系。有时，我们可以借助简单的笔记辅助记忆这些逻辑关系。

3. 流畅表达

表达是翻译的最后一个阶段，也是决定口译质量的关键。一般来说，口译表达要求具有条理性、系统性、科学性和逻辑性，并且同时达到“信、达、雅”的标准。换句话说，就是把原文中的信息，进行有条理地传达，忠实地保持原文的系统性、科学性和逻辑性。在科技英语口译过程中，为了明确地表达出原文的逻辑关系，并且兼顾“信、达、雅”的基本要求，就需要灵活应用翻译中的各种方法，切忌一概应用顺译法。此外，口译人员还要熟练掌握涉及的专业知识，这样才能做到用“行话”进行表达，同时还要注意符合汉语表达习惯。

实训课程三：科技英语口译

Today, industrial electronic systems employ several devices that are described by the term transistor. Each type of transistor has different characteristics and operational conditions that are used to distinguish it from others. In the first part of this discussion, we are concerned with the bipolar junction transistor. Structurally, this transistor is described as bipolar because it has two different current-carrier polarities. Holes are positive current carriers, whereas electrons are negative current carriers. Two distinct kinds of semiconductor crystals are connected together by a common element. The structure of this device is similar to that of two diodes connected back to back, with one crystal being common to both junctions. The center material is usually made thinner than the two outside pieces.

普通英语翻译	科技英语翻译
今天，工业电子系统采用了术语晶体管描述的几种装置。每种晶体管都有不同的特性和操作条件，用以区别于其他晶体管。在本讨论的第一部分中，我们关注的是双极结型晶体管。在结构上，这种晶体管被描述为双极性的，因为它有两种不同的电流载波极性。空穴是正电流载流子，而电子是负电流载流子。两种不同的半导体晶体通过一个公共元件连接在一起。这种装置的结构类似于两个相连的二极管，两个连接处有一个晶体。中心材料通常比两个外件薄。	目前，工业电子系统使用了多种由术语晶体管描述的器件。每种晶体管都有区别于其他种类的特征和工作环境。首先，我们探讨双极型晶体管。从结构上看，这种晶体管被称为双极是因为它具有两种不同的电流载流子极性。空穴是正的电流载流子，电子是负的电流载流子。两种截然不同的半导体晶体通过公共元件连接在一起。该器件的结构类似于两个背靠背连接的二极管，一个晶体同时属于两个结构。其中间的材料通常制造得比外边的两层薄。

对比上面两段译文，就不难看出科技英语和普通英语的区别。科技英语注重的是准确、规范、简练、忠于原文。例如，devices 在原文中形容的是晶体管，因此，考虑实际情况，将其翻译成“器件”比“装置”要更加准确。In the first part of this discussion 原本直译为“在本讨论的第一部分中”，但是这一表达不符合汉语表达习惯，应该意译为“首先”更为贴切。whereas 一词尽管有转折的意思，但在原文中转折的意思并不明显，所以可以不翻译。

科技英语力求反映科技人员客观的叙述和态度，不带主观色彩。科技英语习惯用的语法结构和句型也会体现出客观、准确、严谨的特点。因此，在进行科技口译时，我们要注重理解原文的逻辑关系，忠实地传递原文信息，贴切、准确地选择词义，同时在表达上兼顾“信、达、雅”。

Chapter 4

ELECTRICAL ENGINEERING

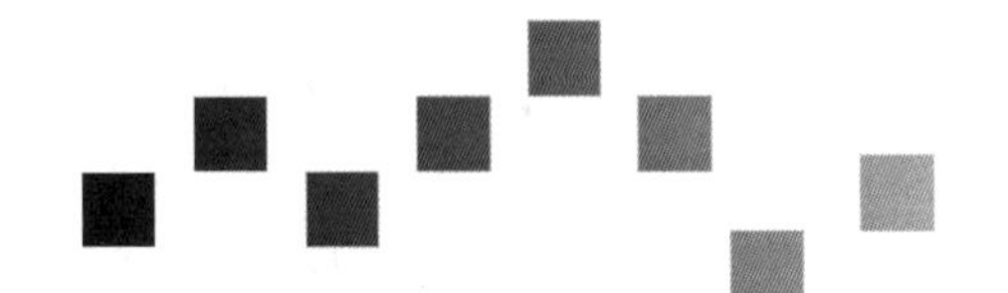

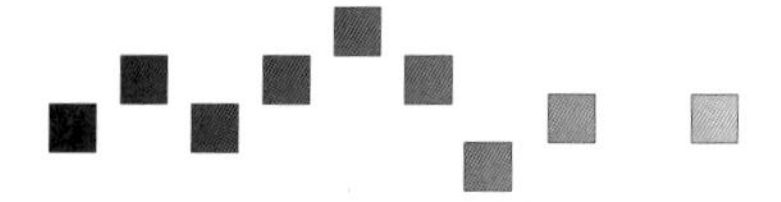

Unit 1 Electric Arc and Static Characteristic

1.1 Electric Arc

The name "arc" was first used by Davy to describe the "arched column of horizontal electrical discharge in air between two charcoal electrodes". Since then, it might have been more logical to confine the term "arc" to the column only. However , the term "arc" is commonly used in place of arched column.

As described earlier, when an electrically conductive channel acquires the plasma state, it converts itself into an arc. The electrodes and their configuration do not only affect the characteristic of an arc but also the type of voltage, source, impedance, the media and their conditions, and so on. It is difficult to classify arc. However, three types of electrical arc discharges, which commonly occur in electrical power system under different conditions, arc distinguished as follows:

(1) free burning arc column in air, a breakdown in very long air gaps, for example ,"lighting discharge".

(2) arc columns in air affected by electrode plasma jet phenomenon in relatively smaller electrode gap distances, for example, spark over on insulator string bushings and isolators in air.

(3) arc columns produced due to high short circuit currents in power systems, for example, "high current arc" in circuit breakers.

Atmospheric air is the most widely used insulation. Under transient conditions with lighting and switching impulses, air insulation systems are designed to have spark over between two electrodes known as protective gaps above a certain level of these voltages. In order to know about the behavior of arc, it is important to learn their characteristic under various conditions.

1.2 Static Voltage-Current (U-I) Characteristic of Arc in Air

Many physicists under various conditions and through different media have measured average arc voltage-current characteristic, with the increasing electrode gap as a parameter, for free burning arc column in steady state. This is a very important characteristic and has received maximum attention in all the arc discharge experiments over a century. King in 1961 came up with an interesting generalized U-I characteristic of arc over a large current range extending from 10^{-4} to 10^{4} A, as shown in Fig. 4-1 (omitted).

Results of measurements made over a number of years in Electrical Research Association Laboratories, Leatherhead, Surrey, England have been collated under various conditions to give consistent characteristic. Similar curves measured in liquid nitrogen by Inaba in 1975 are given.

As seen in Fig. 4-1 (omitted), no sudden transition occurs in the voltage gradient over the entire current range. This curve can be divided into the following three main regions on the basis of accompanied current in arc:

A: below 0.1A

B: 0.1 to 100A

C: above 100A

Region A represents the electrical spark breakdown or the flashover with an unstable leader turning into an arc column. Breakdown with I_i voltage is a typical example for this region. Acceleration of charged particles under the influence of the applied electric field continues to determine the main ionization process, the impact ionization in this region. The gas temperature acquired is, therefore, relatively low, increasing power input, for example, with AC, DC or I_i voltages may gradually raise the gas temperature at particular locations in the stable PB or corona process, which precede the spark breakdown. Thermal ionization may supervene at stem and stable leader corona formation. The breakdown is ultimately accomplished with an unstable leader followed by an arc column. As seen in Fig. 4-1 (omitted), the potential gradient across the discharge column in this region is relatively high since it involves very low magnitudes of current.

From the current magnitude of above 0.1 A (region B), the normal “negative” or “fall” characteristic of the arc discharge begins, lasting almost up to 100 A. From this characteristic it can be safely interpreted that the cross sectional area of the arc column increases with increasing current, resulting in an increase in its temperature. The increase in arc column temperature aggravates thermal ionization, producing more charged particles, electrons and ions, resulting in higher electrical conductivity of the gas, and thus lower potential gradient across it.

An analytical relationship between the potential gradient and the current in the arc column is described in detail in the specialized literature on this subject. This relationship for free burning arc in air, under a steady state within the negative characteristic region B, is described in short in the following paragraphs.

Assuming a uniform distribution of the gas temperature and also the density over the total cross sectional area of an arc, and neglecting the current caused by the movement of ions, the following equation relating arc current density (J) and the potential gradient (E) is obtained:

$$J=\pi r^2 v / Nq=\pi r^2 k_t ENq \tag{4-1}$$

where

r = radius of arc column in cm

v = velocity of electrons in cm/s

N = number of electrons per cm^3

q = electron charge in C

and k_t= movability of electrons in $\left[\frac{\text{cm/s}}{\text{V/cm}}\right]$

Since N and k_t are dependent upon the gas temperature (T) of the arc column, the above equation can be written again as follows:

$$J=\text{r}^2 Ef(T) \tag{4-2}$$

Under the steady state condition of an arc in the region B of Fig. 4-1(omitted), often a generalized relationship between J and E is given as

$$E=\text{const}\,J^{-n}=CJ^{-n} \tag{4-3}$$

Under the assumption that the radiation of heat from the arc is proportional to the column radius or the surface area, the theoretical value of n is calculated to be equal to 0.333. However, according to the investigations made by different research works, its practical value varies from 0.5 to 0.25. The value of n is taken to be equal to 0.5 for low arc currents up to 10 A. For higher current, it may be taken between 0.35 and 0.25, which is close to the theoretical value. Eq. (4-2) holds good also for the arc in air fed by power frequency for the maximum current magnitudes, for example, in isolators. Value of the constant C depends upon the length of the arc; thus, all the factors affecting the arc length, for example, the electrode shape and the starting current of the arc, and so on, affect the value of this constant. The value of C equal to 100 cm/VA may be taken for the maximum length of an arc under given condition.

The current range above 100 A in region C, Fig. 4-1 (omitted) invokes a number of new features. As the current is increased, the so called "plasma jets" and self-generated gas blasts created by these jets are associated with the anode and the cathode extending into the arc column for a distance up to 5 to 6 mm. The plasma jets become prominent in region *C*, rendering it difficult to keep the arc free from electrode vapor. In order to obtain true characteristic of the arc in this current range, experiments must be performed on longer gaps (a few cm to m) to be able to safely assume the plasma jet mechanism to have confined integrally within the electrode. Because of the electrode effect, a definite positive characteristic results in the higher current region. If the gap length is increased, a contraction in the column in the electrode region is affected. Thus, a reasonably flat characteristic is obtained. A fairly constant voltage gradient is measured on a 30 cm gap within a current range of 100 to 1000 A as shown in Fig. 4-1 (omitted). This is the gap length region, in which circuit breakers operate, breaking very high short circuit currents in kA range. The region of minimum voltage gradient of 10 V/cm extends with increasing current for long gap distance. This indicates that for very long arc in air, the value of 10 V/cm can be accepted from 100 to 10.000 A current range. The longitudinal potential gradient in free burning arc in air can be safely taken from this curve to vary between 10 and 50 V/cm for arc currents beginning from 1 A to the highest possible current on arc.

Specialized English Words

arc 电弧
characteristic 特性；特征
electrode 电极
impedance 阻抗
air gap 气隙
insulation 绝缘
isolator 隔离器
transition 转换；变换
corona 电晕
static 静态
horizontal 水平的
conductive 导电的
discharge 放电
plasma 等离子
insulator 绝缘体
breaker 断路；开关
ionization 电离
gradient 斜率；梯度

Unit 2 Dynamic Characteristic of Arc and Extinction Arc

2.1 Dynamic Characteristic of Arc

In case in which the voltage applied is not steady but changing with time, the magnitude of current and hence the plasma characteristic of an arc depend upon the instantaneous voltage. This characteristic of arc is accomplished with a time lag, known as the "thermal time constant". An extremely rapid changing current (due to the transient voltage) through the arc is not able to appreciably affect its temperature within a short time. Hence the arc can be considered to offer constant resistance during this time period. Fig. 4-2 (omitted) shows a schematic of U-I characteristic of arc for different rates of change of current with time. Beginning with a current having no change $\frac{\mathrm{d}i}{\mathrm{d}t}\to 0$ (with time static characteristic), to a current changing very rapidly $\frac{\mathrm{d}u}{\mathrm{d}t}\to\infty$ (with constant resistance characteristic), the predicted characteristic is shown in Fig. 4-2 (omitted).

Consider a power frequency source feeding an arc over a resistance Rs as shown on Fig. 4-2 (a) (omitted). The instantaneous value of voltage variation in the arc is shown in Fig. 4-2 (b) (omitted). When the applied voltage passes through its natural zero, the voltage and, therefore, the current through the arc changes quite rapidly as the rate of change of applied voltage is very high in this region. Near the peak of the applied voltage, where the rate of its change is slow, the voltage and the current in the arc also do not change much with time. Accordingly, the area of cross section of the arc column changes in synch with the instantaneous values of the current. When the current approaches zero, the arc column becomes very thin and it even extinguishes completely for a very short duration (of the order of a few *u*s).

Because of the thermal time constant of the arc column, a hysteresis loop is measured in such a U-I characteristic as shown in Fig. 4-2 (c) (omitted), for 50 Hz current, a relatively large arc time constant, of the order of ms, develops. Such discharge in free air may take place due to faults in the power system. These may also be caused due to the 50 Hz follow-up current between two electrodes, for example, in isolators.

For a small peak arc current as shown in Fig. 4-2 (omitted), as the current increases, a fall in voltage takes place and the area covered by the hysteresis loop is small. On the contrary, for large current magnitude, as the current increases, a rising characteristic is measured and the area covered by the hysteresis loop is also large.

In Gas Insulated Systems (GIS) and in vacuum circuit breakers up to a few tens of re-strikes of current may take place in a single switching operation. This happens because the arc is not quenched and it restrikes several times before the final interruption of the current in these disconnector switches. Since the disconnector or contacts in GIS and vacuum circuit breakers move slowly and the dielectric strength of the medium in these cases is quite high, a number of restrikes occur during opening operation of the switches. The time interval between the restrikes increases with the contact gap distance while opening. Each restrike of the arc generates "very fast transient overvoltage" (VFTO), having different levels of magnitude, of the order of up to 3.0 pu. The rise time of VFTO can be as small as 10 ns. Since the restrikes of the arc are caused in turn by the rapidly changing transient overvoltages, the arc thus formed can be considered to offer constant resistance during the operation period of the switch, as shown by the dynamic characteristic of arc.

2.2 Extinction Arc

An arc in the free air is affected by the wind, its speed direction, the dispersion of heat and even the magnetic forces developed due to its own current. The wind conditions are unpredictable, whereas the dispersion of heat is mainly in an upward direction. The electromagnetic forces, developed due to the magnetic field of its own current, can be directed to act favorably and help the extinction of the arc. These forces are developed by design so that a minimum magnetic induction force is required to drift the arc away from its feeding point. This is an important aspect for consideration while designing the disconnector switch electrode system between which arc may appear in the power system.

An arc in the free air caused by a transient overvoltage in the power system is extinguished when the overvoltage passes away in the form of a travelling wave. But to protect the power system from such faults and power frequency follow-up currents, the applied voltage must be tripped. This is taken care of mostly by the auto-reclosing schemes provided with the power system protection.

Rush cooling of arc column and high rate of rise of dielectric strength of the space between the electrodes in arc extinction chambers of circuit breakers is achieved by injecting fresh air gas or oil. However the techniques of quenching the arc caused by heavy short circuit current in circuit breakers is itself a specialized subject and it is not within the scope of this book.

Specialized English Words

dynamic 动态
instantaneous voltage 瞬时电压
hysteresis loop 磁滞环路
disconnector 断路器；隔离开关
dispersion 色散；散射
extinction arc 灭弧
transient 瞬变
vacuum 真空
dielectric 电介质
electromagnetic forces 电磁力

Unit 3 High Voltage

3.1 Power Transformers

The power transformer is a major power system component that permits economical power transmission with high efficiency and low series voltage drops. Since electric power is proportional to the product of voltage as current, low current levels (and therefore low I^2R losses and low IZ voltage drops) can be maintained for given power levels via high voltages. Power transformers transform AC voltage and current to optimum levels for generation, transmission, distribution, and utilization of electric power.

The development in 1885 by William Stanley of a commercially practical transformer was what made AC power systems more attractive than DC power systems. The AC system with a transformer overcame voltage problems encountered in DC systems as load levels and transmission distances increase. Today's modem power transformers have nearly 100% efficiency, with ratings up to and beyond 1300 MV·A.

In this unit, we discuss the four basic transmission line parameters: series resistance, series inductance, shunt capacitance, and shunt conductance. We also investigate transmission line electric and magnetic fields.

Series resistance accounts for ohmic (I^2R) line losses. Series impedance including resistance and inductive reactance, gives rise to series voltage drop along the line. Shunt capacitance gives rise to line-charging currents. Shunt conductance accounts for U^2G line losses due to leakage currents between conductors or between conductors and ground. Shunt conductance of overhead lines is usually neglected.

3.2 Transmission Lines: Steady State Operation

In this unit, we analyze the performance of single phase and balanced three phase transmission lines under normal steady state operating conditions. Expressions for voltage and current at any point along a line are developed, where the distributed nature of the series impedance and shunt admittance is taken into account. A line is treated here as a two-port network for which the ABCD parameters and an equivalent Πcircuit are derived. Also, approximations are given for a medium length line lumping the shunt admittance, for a short line neglecting the shunt admittance, and for a lossless line assuming zero series resistance and shunt conductance. The concepts of surge impedance loading and transmission-line wavelength are also presented.

3.3 Power Flows

Successful power system operation under normal balanced three phase steady state conditions requires the following:

(1) Generation supplies the demand (load) plus losses.

(2) Bus voltage magnitudes remain close to rated values.

(3) Generators operate within specified real and reactive power limits.

(4) Transmission line and transformers are not overloaded.

The power flow computer program (sometimes called load flow) is the basic tool for investigating these requirements. This program computes the voltage magnitude and angle at each bus in a power system under balanced three phase steady state conditions. It also computes real and reactive power flows for all equipment interconnecting the buses, as well as equipment losses.

Both existing power systems and proposed changes including new generation and transmission to meet projected load growth are of interest.

Conventional nodal or loop analysis is not suitable for power studies because the input data for loads are normally given in terms of power rather than impedance. Also, generators are considered its power sources or current sources. The power flow problem is therefore formulated as a set of nonlinear algebraic equations suitable for computer calculation.

3.4 Transmission Lines: Transient Operation

Transient overvoltages caused by lightning strikes to transmission lines and by switching operations are of fundamental importance in selecting equipment insulation levels and surge protection devices. We must, therefore, understand the nature of transmission line transients.

When a line with distributed constants is subjected to a disturbance such as a lightning strike or a switching operation, voltage and current waves arise and travel along the line at a velocity near the speed of light. When these waves arrive at the line terminals, reflected voltage and current waves arise and travel back down the initial waves.

Because of line losses, traveling waves are attenuated and essentially die out after a few reflections. Also, the series inductances of transformer windings effectively block the disturbances, thereby preventing them from entering generator windings. However, due to the reinforcing action of several reflected waves, it is possible for voltage to build up to a level that could cause transformer insulation or line insulation to arc over and suffer damage.

Circuit breakers, which can operate within 50 ms, are too slow to protect against lightning or switching surges. Lightning surges can rise to peak levels within a few microseconds and switching surges within a few hundred microseconds can be fast enough to destroy insulation before a circuit breaker could open. However, protective devices are available. Surge arresters can be used to protect equipment insulation against transient overvoltages. These devices limit voltage to a ceiling level and absorb the energy from lightning and switching surges.

Specialized English Words

power transformer 电力变压器
voltage drop 电压降
distribution 配电；分布
inductance 电感
impedance 阻抗
reactance 电抗
single-phase 单相
transmission 传输
generation 生成；产生
utilization 利用；使用
capacitance 电容
conductance 电导
performance 性能
admittance 导纳

Unit 4 The Nature of Relaying

The function of protective relaying is to promptly remove from service any element of the power system that starts to operate in an abnormal manner. In general, relays do not prevent damage to equipment: they operate after some detectable damage has already occurred. Their purpose is to limit, to the extent possible, further damage to equipment, to minimize danger to people, to reduce stress on other equipment and, above all, to remove the faulted equipment from the power system as quickly as possible so that the integrity and stability of the remaining system is maintained. The control aspect of relaying systems also helps return the power system to an acceptable configuration as soon as possible so that service to customers can be restored.

4.1 Reliability, Dependability and Security

Reliability is generally understood to measure the degree of certainty that a piece of equipment will perform as intended. Relays, in contrast with most other equipment, have two alternative ways in which they can be unreliable: they may fail to operate when they are expected to, or they may operate when they are not expected to. This leads to a two-pronged definition of reliability of relaying systems: a reliable relaying system must be dependable and secure. Dependability is defined as the measure of the certainty that the relays will operate correctly for all the faults for which they are designed to operate. Security is defined as the measure of the certainty that the relays will not operate incorrectly for any fault.

4.2 Selectivity of Relays and Zones of Protection

The property of security of relays, that is, the requirement that they not operate for faults for which they are not designed to operate, is defined in terms of regions of a power system called zones of protection for which a given relay or protective system is responsible. The relay will be considered to be secure if it responds only to faults within its zone of protection. Relays usually have inputs from several current transformers (CTs), and the zone of protection is bounded by these CTs. The CTs provide a window through which the associated relays "see" the power system inside the zone of protection. While the CTs provide the ability to detect a fault inside the zone of protection, the circuit breakers (CBs) provide the ability to isolate the fault by disconnecting all of the power equipment inside the zone. Thus, a zone boundary is usually defined by a CT and a CB. When the CT is a part of the CB, it becomes a natural zone boundary. When the CT is not an integral part of the CB, special attention must be paid to the fault detection and fault interruption logic. The CT still defines the zone of protection, but communication channels must be used to implement the tripping function from appropriate remote locations where the CBs may be located.

In order to cover all power equipment by protection systems, the zones of protection must meet the following requirements:

(1) All power system elements must be encompassed by at least one zone. Good relaying practice is to be sure that the more important elements are included in at least two zones.

(2) Zones of protection must overlap to prevent any system element from being unprotected. Without such an overlap, the boundary between two nonoverlapping zones may go unprotected. The region of overlap must be finite but small, so that the likelihood of a fault occurring inside the region of overlap is minimized. Such faults will cause the protection belonging to both zones to operate, thus removing a larger segment of the power system from service.

4.3 Relay Speed

It is, of course, desirable to remove a fault from the power system as quickly as possible. However, the relay must make its decision based upon voltage and current waveforms which are severely distorted due to transient phenomena which must follow the occurrence of a fault. The relay must separate the meaningful and significant information contained in these waveforms upon which a secure relaying decision must be based. These considerations demand that the relay take a certain amount of time to arrive at a decision with the necessary degree of certainty. The relationship between the relay response time and its degree of certainty is an inverse one, and this inverse time operating characteristic of relays is one of the most basic properties of all protection systems.

Although the operating time of relays often varies between wide limits, relays are generally classified by their speed of operation as follows:

(1) Instantaneous. These relays operate as soon as a secure decision is made. No intentional time delay is introduced to slow down the relay response.

(2) Time delay. An intentional time delay is inserted between the relay decision time and the initiation of the trip action.

(3) High speed. These relays operate in less than a specified time. The specified time in present practice is 50 ms (3 cycles on a 60 Hz system).

(4) Ultra high speed. This term is not included in the *Relay Standards* but is commonly considered to be operation in 4 ms or less.

4.4 Primary and Backup Protection

A protection system may fail to operate and, as a result, fail to clear a fault. It is thus essential that provision be made to clear the fault by some alternative protection system or systems. These alternative protection system(s) are referred to as duplicate, backup or breaker-failure protection systems. The main protection system for a given zone of protection is called the primary protection system. It operates in the fastest time possible and removes the least amount of equipment from service. On EHV systems, it is common to use duplicate primary protection systems in case an element in one primary protection chain may fail to operate. This duplication is therefore intended to cover the failure of the relays themselves. One may use relays from a different manufacturer, or relays based upon a different principle of operation, so that some inadequacy in the design of one of the primary relays is not repeated in the duplicate system. The operating times of the primary and the duplicate systems are the same.

4.5 Single and Three Phase Tripping and Reclosing

As a large proportion of faults on a power system are of a temporary nature, the power system can be returned to its prefault state if the tripped circuit breakers are reclosed as soon as possible. Reclosing can be manual. That is, it is initiated by an operator working from the switching device itself, from a control panel in the substation control house or from a remote system control center through a supervisory control and data acquisition (SCADA) system. Clearly, manual reclosing is too slow for the purpose of restoring the power system to its prefault state when the system is in danger of becoming unstable. Automatic reclosing of circuit breakers is initiated by dedicated relays for each switching device, or it may be controlled from a substation or central reclosing computer. All reclosing operations should be supervised (i.e., controlled) by appropriate interlocks to prevent an unsafe, damaging or undesirable reclosing operation. Some of the common interlocks for reclosing are the following:

(1) Voltage check. Used when good operating practice demands that a certain piece of equipment be energized from a specific side. For example, it may be desirable to always energize a transformer from its high voltage side. Thus if a reclosing operation is likely to energize that transformer, it would be well to check that the circuit breaker on the low voltage side is closed only if the transformer is already energized.

(2) Synchronizing check. This check may be used when the reclosing operation is likely to energize a piece of equipment from both sides. In such a case, it may be desirable to check that the two sources which would be connected by the reclosing breaker are in synchronism and approximately in phase with each other. If the two systems are already in synchronism, it would be sufficient to check that the phase angle difference between the two sources is within certain specified limits. If the two systems are likely to be unsynchronized, and the closing of the circuit breaker is going to synchronize the two systems, it is necessary to monitor the phasors of the voltages on the two sides of the reclosing circuit breaker and close the breaker as the phasors approach each other.

(3) Equipment check. This check is to ensure that some piece of equipment is not energized inadvertently.

Specialized English Words

relay 继电器；延迟
abnormal 反常的，异常的
equipment 装置；设备
stability 稳定性
configuration 配置
restore 再存入；还原
reliability 可靠性
dependability 可信任；可靠性
property 特性
zones of protection 防护区域
circuit breakers (CBs) 断路器
current transformers (CTs) 电流互感器
waveform 波形
transient 瞬变
instantaneous value 瞬时值
time delay 延时
backup 备存
synchronize 使同步；使同时
supervisory control and data acquisition (SCADA) 监控与数据采集

Unit 5 Detection of Faults of Power System Relay Protection

5.1 Introduction

Because the purpose of power system protection is to detect faults or abnormal operating conditions, relays must be able to evaluate a wide variety of parameters to establish that corrective action is required. The most common parameters that reflect the presence of a fault are the voltages and currents at the terminals of the protected apparatus or at the appropriate zone boundaries. Occasionally, the relay inputs may also include states—open or closed—of some contacts or switches. A specific relay, or a protection system, must use the appropriate inputs, process the input signals, and determine that a problem exists, and then initiate some action. In general, a relay can be designed to respond to any observable parameter or effect. The fundamental problem in power system protection is to define the quantities that can differentiate between normal and abnormal conditions. This problem of being able to distinguish between normal and abnormal conditions is compounded by the fact that "normal" in the present sense means that the disturbance is outside the zone of protection. This aspect, which is of the greatest significance in designing a secure relaying system, dominates the design of all protection systems. For example, consider the relay shown in Fig. 4-3 (omitted). If one were to use the magnitude of a fault current to determine whether some action should be taken, it is clear that a fault on the inside (fault F_1), or on the line outside (fault F_2), of the zone of protection is electrically the same fault, and it would impossible to tell the two faults apart based upon the current magnitude alone. Much ingenuity is needed to design relays and protection systems that would be reliable under all the variations to which they are subjected throughout their life.

Whether, and how, a relaying goal is met is dictated by the power system and the transient phenomena it generates following a disturbance. Once it is clear that a relaying task can be performed, the job of designing the hardware to perform the task can be initiated. The field of relaying is almost 100 years old. Ideas on how relaying should be done have evolved over this long period, and the limitations of the relaying process are well understood. As time has gone, the hardware technology used in building the relays has gone through several major changes: relays began as electromechanical devices, then progressed to solid state hardware in the late 1950s, and more recently they are being implemented on microcomputers. We will now examine in general terms the functional operating principles of relays and certain of their design aspects.

5.2 Detection of Faults

In general, as faults (short circuits) occur, currents increase in magnitude, and voltages go down. Besides these magnitude changes of the AC quantities, other changes may occur in one or more of the following parameters: phase angles of current and voltage phasors, harmonic components, active and reactive power, frequency of power system, and so on. Relay operating principles may be based upon detecting these changes, and identifying the changes with the possibility that a fault may exist inside its assigned zone of protection. We will divide relays into categories based upon which of these input quantities a particular relay responds.

1. Level Detection

This is the simplest of all relay operating principles. As indicated above, fault current magnitudes are almost always greater than the normal load currents that exist in a power system. Consider the motor connected to a 4 kV power system as shown in Fig. 4-4 (omitted). The full load current for the motor is 245 A. Allowing for an emergency overload capability of 25%, a current of 1.25×245=306 A or lower should correspond to normal operation. Any current above a set level (chosen to be above 306 A by a safety margin in the present example) may be taken to mean that a fault, or some other abnormal condition, exists inside the zone of protection of the motor. The relay should be designed to operate and trip the circuit breaker for all current above the setting, or if desired, the relay may be connected to sound an alarm, so that an operator can intervene and trip the circuit breaker manually or take other appropriate action.

The level above which relay operates is known as the pickup setting of the relay. For all currents above the pickup, the relay operates, and for currents smaller than the pickup value, the relay takes no action. It is of course possible to arrange the relay to operate for values smaller than the pickup value, and take no action for values above the pickup. An undervoltage relay is an example of such a relay.

The operating characteristics of an overcurrent relay can be presented as a plot of the operating time of the relay versus the current in the relay. It is best to normalize the current as a ratio of the actual current to the pickup setting. The operating time for (normalized) currents less than 1.0 is infinite, while for values greater than 1.0 the relay operates. The actual time for operation will depend upon the design of the relay. The ideal level detector relay would have a characteristic as shown by the solid line in Fig. 4-5 (omitted). In practice, the relay characteristic has a less abrupt transition, as shown by the dotted line.

(1) Magnitude Comparison

This operating principle is based upon the comparison of one or more operating quantities with each other. For example, the current balance relay may compare the current in one circuit with the current in another circuit, which should have equal or proportional magnitudes under normal operating conditions.

The relay will operate when the current division in the two circuits varies by a given tolerance. Fig. 4-6 (omitted) shows two identical parallel lines that are connected to the same bus at either end. One could use a magnitude comparison relay that compares the magnitudes of the two line currents, i_A and i_B. If $|i_A|$ is greater than $|i_B| + \in$ (where $\in$ is a suitable tolerance), and line B is not open, the relay would declare a fault on line A and trip it. Similar logic would be used to trip line B if its current exceeds that in line A, when the latter is not open. Another instance in which this relay can be used is when the windings of a machine have two identical parallel subwindings per phase.

(2) Differential Comparison

Differential comparison is one of the most sensitive and effective methods of providing protection against faults. The concept of differential comparison is quite simple, and can be best understood by referring to the generator winding shown in Fig. 4-7 (omitted). As the winding is electrically continuous, current entering one end, I_1, must equal the current leaving the other end, I_2. One could use a magnitude comparison relay described above to best for a fault on the protected winding. When a fault occurs between the two ends, the two currents are no longer equal. Alternatively , one could form an algebraic sum of the two currents entering the protected winding, that is, $(I_1 - I_2)$, and use a level detector relay to detect the presence of a fault. In either case, the protection is termed a differential protection. In general, the differential protection principle is capable of detecting very small magnitudes of fault current. Its only drawback is that it requires currents from the extremities of a zone of protection, which restricts its application to power apparatus, such as transformers, generators, motor, buses, capacitors, and reactors.

(3) Phase Angle Comparison

This type of relay compares the relative phase angle between two AC quantities. Phase angle comparison is commonly used to determine the direction of a current with respect to a reference quantity. For instance, the normal power flow in a given direction will result in the phase angle between the voltage and the current varying around its power factor angle, approximately ±30°.

When the power flows in the opposite direction, this angle will become $(180^\circ \pm 30^\circ)$. Similarly, for a fault in the forward or reverse direction, the phase angle of the current with respect to the voltage will be $-\varphi$ and $(180^{\circ}-\varphi)$, respectively, where φ,the impedance angle of the fault circuit is close to 90° for power transmission networks. These relationships are explained for two transmission lines in Fig. 4-8 (omitted). This difference in phase relationships created by a fault is exploited by making relays that respond no phase angle differences between two input quantities—such as the fault voltage and the fault current in the present example.

2. Distance Measurement

As discussed above, the most positive and reliable type of protection compares the current entering the circuit with the current leaving it. On transmission lines and feeders, the length, voltage, and configuration of the line may make this principle uneconomical. Instead of comparing the local line current with the far-end line current, the relay compares the local current with the local voltage. This, in effect, is a measurement of the impedance of the line as seen from the relay terminal. An impedance relay relies on the fact that the length of the line (i.e., its distance) for a given conductor diameter and spacing determines its impedance.

3. Pilot Relaying

Certain relaying principles are based upon the information obtained by the relay from a remote location. The information is usually, although not always, in the form of contact status (open or closed). The information is sent over a communication channel using power line carrier, microwave, or telephone circuits.

4. Harmonic Content

Currents and voltages in a power system usually have sinusoidal waveform of fundamental power system frequency. There are, however, deviations from a pure sinusoid, such as the third harmonic voltages and currents produced by the generators that are present during normal system operation. Other harmonics occur during abnormal system conditions, such as the odd harmonics associated with transformer saturation, or transient components caused by the energization of transformers. These abnormal conditions can be detected by sensing the harmonic content through filters in electromechanical or solid-state relays, or by calculation in digital relays. Once it is determined that an abnormal condition exists a decision can be made whether some control action is required.

5. Frequency Sensing

Normal power system operation is at 50 or 60 Hz, depending upon the country. Any deviation from these values indicates that a problem exists or is imminent. Frequency can be measured by filter circuits, by counting zero crossings of waveforms in a unit of time, or by special sampling and digital computer techniques. Frequency-sensing relays may be used to take corrective actions that will bring the system frequency back to normal.

The various input quantities described above, upon which fault detection is based, may be used either singly or in any combination, to calculate power, power factor, directionality, impedance, and so on, and can in turn be used as relay-actuating quantities. Some relays are also designed to respond to mechanical devices such as fluid level detectors, pressure, or temperature sensors, and so on. Relays may be constructed from electromechanical elements such us solenoids, hinged armatures, induction discs, solid-state elements such as diodes, silicon controlled rectifiers (SCRs), transistors, or magnetic or operational amplifiers, or digital computers using analog-to-digital converters and microprocessors. It will be seen that, because the electromechanical relays were developed early on in the development of protection systems, the description of all relay characteristics is often in terms of electromechanical relays. The construction of a relay does not inherently change the protection concept, although there are advantages and disadvantages associated with each type. We will examine the various hardware options for relays in the following section.

Specialized English Words

detection 检测
relay 继电器；延迟
contact 接触
harmonic 谐波
reactive (passive) power 无功功率
detector 检测器
microwave 微波
solenoid 螺线管
fault 故障；错误
apparatus 仪器；装置
phasor 相量；相图
active power 有功功率
tolerance 公差；容差
power factor 功率因数
electromechanical 电机的；机电的
microprocessor 微处理器

Unit 6　专业英语翻译

6.1 专业英语的翻译标准与过程

翻译标准是指评价译文质量的尺度，也是指导翻译的准绳。我国影响最为深远的翻译标准是由著名启蒙思想家严复提出的“信、达、雅”。然而，与文学作品不同，科技英语具有结构严谨，逻辑性强，公式、数据、专业术语多等特点。因此，科技英语的翻译对“信、达、雅”的处理更侧重于“忠实、通顺、简练”。所谓忠实，主要是指忠实于原文的技术内容。把原文的内容准确、完整地表达出来，既不篡改、歪曲，又不任意遗漏和删减。所谓通顺，主要是指译文要通顺易懂，要符合汉语的规范和表达习惯，同时也要符合专业表达的要求。所谓简练，主要是指译文要尽可能地简短、精炼，没有冗词废字。一般认为，“忠实、通顺”是专业英语翻译的基本标准，两者相辅相成，互为辩证。

专业英语翻译的过程主要是理解和表达的过程，大致可分为理解、表达、校核三个阶段。其中理解是前提，先要通读原文，领略全文大意，明辨语法形式，弄清句子成分之间的关系，结合上下文推敲词意；表达是关键，初译时以忠实为主，在深刻理解原文的基础之上，考虑汉语的语言习惯，准确选词，合理搭配词的顺序，恰当造句，力求简练；校核就是检查译文的错漏之处，检查译文的前后关系，保持专业英语的逻辑性，同时润色文字，做到翻译的“雅”。

6.2 科技翻译的方法

1. 直译和意译

直译是基本保持原文表达形式及内容的翻译方法，而意译是指保留原文内容、不保持原文形式的翻译方法。直译所强调的是“形似”，主张将原文内容按照原文的形式（包括词序、语序、语气、结构、修辞方法等）用译语表述出来。

直译与意译不是两种完全孤立的翻译方法，两者相互关联、互为补充。在翻译实践中，不应该完全拘泥于其中的一种，必须学会何时采用直译、何时采用意译，运用直译与意译的相关技巧，将直译和意译有机地结合起来，以忠实、通顺地表达出原文的意义为翻译的最终目的。

2. 合译和分译

合译是把原文两个或两个以上的简单句或复合句，译成汉语的一个句子来表达，译文符合汉语的表达习惯。分译是把原文的某个成分，比如一个词、词组或短语，译成汉语的一个句子，使原文的一个句子被译成汉语的两个或两个以上的句子。

由于英汉两种语言的句型结构、修辞手法和表达习惯的差异，在科技英语翻译过程中，为了使译文准确、通顺并且符合汉语的表达习惯，有时需要改变原文句子的结构，合译法和分译法就是改变原文句子结构的两种常用方法。例如，经常采用分译法把原文句子中复杂的逻辑关系表述清楚；经常采用合译法省掉一些重复的词语，使译文更加精炼，符合汉语表达习惯。

3. 增译和省译

增译是翻译时在译文中增添原文省略或无其词而有其义的词语或短句，以便更准确、完整地表达原文包含的意义，符合汉语的表达习惯。

省译是与增译相对应的一种翻译方法，即在翻译过程中，省略某些词不翻译，使得译文简洁明了，符合汉语的思维习惯和语言习惯，避免译文累赘、不顺。

在翻译过程中，译者应遵循汉语的习惯表达方式，在忠实于原文的基础上，适当地进行增译或省译。

汉语和英语在名词、代词、连词、介词和冠词上的使用习惯不同。例如英语中句子一般都要有主语，代词、连词使用频率较高，因此，在汉译英时需要增补主语、代词和连词，而在英译汉时则需要根据情况适当地删减。无论是增译或是省译，在翻译过程中都要以完整、准确、符合汉语的表达习惯为原则。

4. 顺译和倒译

顺译是按照原文的字面意义或语序进行翻译的方法。在翻译时，通常不会使用完全顺译法，即逐字顺译，而是增补或删减句子中的某些词语，调整个别语序，使译文表达更为准确和通顺，即基本顺译法。

倒译是与顺译相对应的一种翻译方法，即在翻译过程中，不按照原文的语序进行翻译的方法。汉语和英语的表达习惯不同。英语在表达时间、动作、结果、条件时，语法手段较多，可以先述也可以后述。当英语和汉语习惯使用的语序不同时，就可以采用倒译法翻译。

以上介绍的各种翻译方法必须灵活应用，无论是采用哪一种方法，其目的都是为了能准确、通顺、简练地用译文表达原文的思想内容。事实上，在科技英语翻译实践中，我们不会完全单独地使用某一种翻译方法，而是需要将各种译法融会贯通、有机结合、灵活使用，以使译文达到理想的翻译标准：准确规范、通顺易懂、简洁明晰。

实训课程四：产品说明书的翻译

产品说明书是生产厂家向消费者介绍产品性质、性能、结构、用途、规格、使用方法、注意事项等时使用的经济应用文书，又称商品说明书、产品说明或说明书，其具有独特文体和一定的格式。要能正确理解原文，翻译通顺，首先要了解和熟悉它的词汇句法的特点，并掌握一定的翻译技巧。

1. 产品说明书的词汇和句法特点

产品说明书在词汇应用和句子结构上有以下特点：词汇应用的专业性和准确性、句子结构的简明性和直观性，以及事物描述的实用性和客观性。

（1）广泛使用复合名词结构。

产品说明书中经常使用复合名词结构代替后置定语，同时还经常使用名词化结构作主语，以求行文简洁、明了、客观。

（2）普遍使用一般现在时。

一般现在时可以用来表示不受时间、空间限制的客观存在，包括客观真理、科学事实等。产品说明书的主要部分是对于产品的功能、用途等的一般叙述，是客观的事实，用一般现在时可以体现出其内容的客观性和形式的简明性。

（3）经常使用被动语态。

产品说明书的主要目的是说明所介绍的产品的客观事实，强调的是所叙述的事物本身，而不需要过多地注意它的行为主体。被动语态的使用使读者的注意力主要集中于受动者，也就是产品本身。

（4）广泛使用祈使句。

产品说明书的主要目的是指导使用者如何来使用它，即告诉使用者要做什么、不要做什么或该怎么做，所以文中经常使用祈使句，旨在指导阅读者如何进行实际操作。使用祈使句使得表述准确、客观、简洁、明了。

2. 产品说明书的音译技巧

产品说明书基本是说明产品的使用方法和操作步骤等，一般不使用文学作品中的比喻、拟人等修辞手法。因此，翻译时以直译为主，意译为辅。

（1）直译。

在翻译英文产品说明书时，直译是最常用的技巧。

（2）意译。

在翻译产品说明书时，有时原文中的一些词语或句子成分必须作适当调整，才能使句子更好地符合汉语表达习惯，这时就需要使用意译的翻译技巧。意译的方法通常有：对原文句子语序的调整、对原文单词词性的转换和对原文词语的增译或减译。

3. 说明书常用句型和译法

例 1：Place the new pump in position and tighten the mounting bolts.

使水泵就位，拧紧装配螺栓。

例 2：Make sure hoses are clamped tightly.

务必将皮管夹紧。

例 3：Remove the AV supply lead before servicing or cleaning heads, rollers etc.

切断交流电源才能维修或清洗磁头、压轮等部件。

例 4：The Voltage Selector Setting should be checked to see if it conforms to the local AC supply voltage.

必须查看本机交流电压选择器之预调状态是否符合本地交流电压。

例 5：To obtain the best performance and ensure years of trouble-free use, please read this instruction manual carefully.

请仔细阅读说明书，以便使本机发挥最佳性能，经久耐用，不出故障。

Exercise——VD4 断路器的使用说明（节选）

ABB VD4 Instruction Manual: Summary

1.1 General

The vacuum circuit-breaker of type VD4 on with drawable part are intended for indoor installation in air-insulated switchgear of withdrawable design. Their switching capacity is sufficient to handle any conditions arising from switching of equipment and systems components under normal operating conditions, particularly short-circuits, within the parameters of their technical data.

Vacuum circuit-breaker have particular advantages for use in networks where there is a high switching frequency in the working current range and/or where a certain number of short-circuit breaking operations are expected. The vacuum circuit-breakers of the type VD4, designed in column form, are suitable for autoreclosing, and have exceptionally high operating reliability and long life.

Together with this instruction manual, it is essential to consult manual BA 359, Vacuum circuit-breaker type VD4, high current.

1.2 Standards and specifications

1.2.1 Switchgear manufacture

The switchgear complies with the following specifications in accordance with DIN VDE/ the relevant IEC publications:

• VDE 0670, part 1000/IEC 60694

• DIN VDE 0670, part 104, and IEC 62271-100

• DIN VDE 0847, part 4, and IEC 61000-4

1.2.2 Installation and operation

The relevant specifications are to be taken into account during installation and operation, particularly:

• DIN VDE 0101, Power installations exceeding AC 1 kV

• VDE 0105, Operation of electrical installations

• DIN VDE 0141, Earthing systems for special power installations with rated voltages above 1 kV

• Accident prevention regulations issued by the appropriate professional bodies or comparable organisations.

In Germany, these comprise the following safety regulations:

– Health and Safety at Work Standards BGV A1 and BGV A2

• Safety guidelines for auxiliary and operating materials

• Order related details provided by ABB Calor Emag

Chapter 5

AUTOMATION

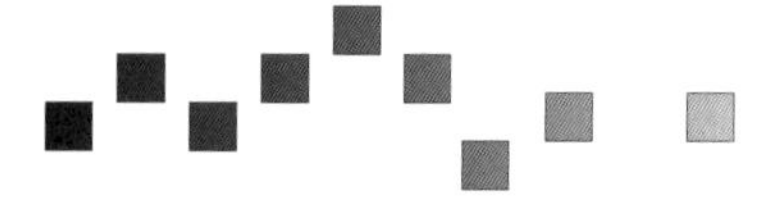

Unit 1 Automation Control Theory

1.1 Introduction

What do we mean by the word control? One of the various dictionary definitions is "a device for regulating a machine". This is a reasonably good definition, although it implies that the "device" is a separate piece of hardware, and the word "regulating" should not be interpreted in its strict technical sense. Perhaps it would be better to say that a control is a scheme (or an algorithm) that makes a machine behave according to our wishes.

There are many quantities that we can control, in the sense that the machines used to manipulate these quantities are operated in such a way that we get what we want. Such quantities, often called the controlled variables, include: temperature, voltage, frequency, flow rate, current, position, horsepower, speed, illumination, altitude (or depth), and many others. The methods used to accomplish the control objectives can be classified, very generally, as open-loop controls and closed-loop or feedback controls.

Open-loop controls are, in general, calibrated systems. They are adjusted to provide the desired result and are expected to duplicate that result as needed because of the adjustment. Fig. 5-1 (a) (omitted) shows a basic block diagram for open-loop control.

Closed-loop feedback controls operate according to a very simple principle:

(1) Measure the variable to be controlled.

(2) Compare this measured value with the desired (commanded) value and determine the difference (error).

(3) Use this difference (amplified if necessary) to adjust the controlled variable so as to reduce the difference (error).

Fig. 5-1 (b) (omitted) illustrates this principle in block diagram form. Examples of feedback control are numerous: Thermostatic control of room temperature, of refrigerator or freezer temperature, oven temperature, etc., all of which measure the temperature, compare it with the desired value as set on the control dial, and turn on or off the heat source (or sink) as needed. Other examples are electric supply voltage and frequency, which are measured at the generating station and compared with desired values. If the voltage is in error, the field current of the generator is adjusted; if the frequency is not correct, the turbine speed is adjusted.

Autopilots on aircraft, submarines, etc., are feedback controllers and satellite tracking antennas are automatic positioning systems. In a computer system, the magnetic disk memory requires a very accurate feedback controller to position the read/write head over the proper memory track.

1.2 When Do We Need Feedback?

From the preceding example it is clear that both open-loop controls and feedback controls are used to control the same kinds of variables. What, then, are the advantages and disadvantages of each? Why do we need feedback control and when should we use it?

In general, the open-loop control is much simpler and less expensive. No sensors are needed to measure the variables and provide the feedback. Also it usually does not require much amplification. However, it is not very accurate (compared with feedback control), and its accuracy can vary without this variation being detected by the system.

For example, consider a sprinkler used to water the lawn. The system is adjusted to water a given area by opening the water valve and observing the resulting pattern. When the pattern is considered satisfactory, the system is "calibrated" and no further valve adjustment is made. The pattern will be maintained reasonably well if there is no change in water pressure. When a tap is opened inside the house, reducing pressure, the pattern changes, i.e., the open-loop control has no means of maintaining the desired steady-state condition.

Inspection of the previously listed examples indicates, correctly, that most automatic controls are used primarily to maintain the controlled variable at a fixed steady-state value. However, most systems are subject to disturbances that perturb the controlled variable, and a major reason for the use of feedback control is to reduce the effect of these disturbances. For example, when there is an environmental change such as an increase in ambient temperature from 40° at 6 a.m. to 75° at noon, an open-loop house heating system is not satisfactory, but a thermostatically controlled heater can maintain the desired room temperature. In most systems, the desired values of the variable may be changed (we often change the temperature setting on our room thermostat). When the controlled variable is disturbed, the recovery time (settling time) is often slow, and in most cases, not much can be done to speed it up.

The closed-loop system is inherently more accurate than the open-loop system because of its principle of operation. In addition, it can be designed to provide extreme accuracy in the steady state (for almost all steady-state operating conditions). For example, in modem magnetic disk memories, the width of a memory track may be less than 0.002 in, and the positioning controller must align the read/write head over the track center with an error of less than 10 uin. Such accuracy is not possible with an open-loop controller.

The closed-loop system has an additional advantage in that its response time can be adjusted by appropriate design. Normally the response time can be reduced substantially below that available from an equivalent open-loop system.

Unfortunately, the feedback link introduces a serious problem-stability! Feedback systems readily become oscillators: that is, the controlled variable fluctuates continuously at some periodic rate (frequency) and never reaches the desired steady-state condition. For any specific system, stable operation can be achieved by proper design, but the design problem is not always an easy one to solve.

1.3 Basic Procedures

When we wish to develop a feedback control system for a specific purpose, the general procedure may be summarized as follows:

(1) Choose a way to adjust the variable to be controlled: e.g., the mechanical load will be positioned with an electric motor or the temperature will be controlled by an electrical resistance heater.

(2) Select suitable sensors, power supplies, amplifiers, etc., to complete the loop.

(3) Determine what is required for the systems to operate with the specified accuracy in a steady state and for the desired response time.

(4) Analyze the resulting system to determine its stability.

(5) Modify the system to provide stability and other desired operating conditions by redesigning the amplifier/controller, or by introducing additional control loops.

Specialized English Words

automation control　自动控制
frequency　频率
horsepower　马力
calibrated system　校准系统
quantity　量；数量
flow rate　流量；流速
illumination　照明
open-loop control　开环控制

measured value 测量值
commanded value 指令值
steady state 稳态
stability 稳定性
feedback control 反馈控制
desired 期望
difference/error 误差
response time 响应时间
closed-loop 闭环

Unit 2 Control and Estimation of Induction Motor Drives

2.1 Introduction

The control and estimation of induction motor drives constitute a vast subject, and the technology has further advanced in recent years. Induction motor drives with cage-type machines have been the workhorses in industry for variable-speed applications in a wide power range that covers from fractional horsepower to multi-megawatts. These applications include pumps and fans, paper and textile mills, subway and locomotive propulsions, electric and hybrid vehicles, machine tools and robotics, home appliances, heat pumps and air conditioners, rolling mills, wind generation systems, etc. In addition to process control, the energy-saving aspect of variable-frequency drives is getting a lot of attention nowadays.

The control and estimation of ac drives in general are considerably more complex than those of dc drives, and this complexity increases substantially if high performances are demanded. The main reasons for this complexity are the need of variable-frequency, harmonically optimum converter power supplies, the complex dynamics of ac machines, machine parameter variations, and the difficulties of processing feedback signals in the presence of harmonics. While considering drive applications, we need to address the following questions:

① One-, two- or four-quadrant drive?

② Torque, speed, or position control in the primary or outer loop?

③ Single- or multi-motor drive?

④ Range of speed control? Does it include zero speed and field-weakening regions?

⑤ Accuracy and response time?

⑥ Robustness with load torque and parameter variations?

⑦ Control with speed sensor or sensorless control?

⑧ Type of front-end converter?

⑨ Efficiency, cost, reliability, and maintainability considerations?

⑩ Line power supply, harmonics, and power factor considerations?

In this unit, we will study different control techniques of induction motor drives, including scalar control, vector or field-oriented control, direct torque and flux control, and adaptive control. The estimation of feedback signals, particularly speed estimation in sensorless vector controls will be discussed. Variable-frequency power supplies will be considered with voltage-fed inverters, current-fed inverters, and cyclo- converters. However, emphasis will be given to voltage-fed converters because of their popularity in industrial drives.

2.2 Induction Motor Control with Small Signal Model

The general control block diagram for variable-frequency speed control of an induction motor drive is shown in Fig.5-2(ommited). It consists of a converter-machine system with hierarchy of control loops added to it. The converter-machine unit is shown with voltage (V_S^*) and frequency (ω_e^*) as control inputs. The outputs are shown as speed (ω_r), developed torque (T_e), stator current (I_s), and rotor flux (ψ_r). Instead of voltage control, the converter may be current-controlled with direct or indirect voltage control in the inner loop. There are coupling effects between the input and output variables. For example, both the torque and flux of a machine are functions of voltage and frequency. Machine parameters may vary with saturation, temperature, and skin effect, adding further nonlinearity to the machine model. The converter can be described by a simplified model, which consists of an amplifier gain and a dead-time lag due to the PWM sampling delay. The system becomes discrete-time because of the converter and digital control sampling effects. All the control and feedback signals can be considered as dc and proportional to actual variables. The speed control is shown with an inner torque control loop, which may be optional. Adding a high-gain inner loop control provides the advantages of linearization, improved bandwidth, and the ability to control the signals within safe limits. Like a dc machine, the flux of an ac machine is normally controlled to be constant at the rated value because it gives fast response and high developed torque per ampere of current. In fact, the flux under consideration may be stator flux (ψ_s), rotor flux (ψ_r), or air gap flux (ψ_m or ψ_g) . However, the rotor flux control is considered in the present case. The inner control loops have faster response (i.e., higher bandwidth) than the outer loop. The "controller" block shown in the figure may have different structures, which will be discussed later.

Since an ac drive system is multivariable, nonlinear with internal coupling effect, and discrete-time in nature, its stability analysis is very complex. Computer simulation study becomes very useful for investigating the performance of the drive, particularly when a new control strategy is developed. Once the control structure and parameters of the control system are determined by the simulation study for acceptable performance, a prototype system can be designed and tested with further iteration of the controller parameters.

2.3 Scalar Control

Scalar control, as the name indicates, is due to magnitude variation of the control variables only, and disregards the coupling effect in the machine. For example, the voltage of a machine can be controlled to control the flux, and frequency or slip can be controlled to control the torque. However, flux and torque are also functions of frequency and voltage, respectively. Scalar control is in contrast to vector or field-oriented control (will be discussed later), where both the magnitude and phase alignment of vector variables are controlled. Scalar-controlled drives give somewhat inferior performance. but they are easy to implement. Scalar-controlled drives have been widely used in industry. However, their importance has diminished recently because of the superior performance of vector-controlled drives, which is demanded in many applications.

2.4 Vector or Field-oriented Control

So far, we have discussed scalar control techniques of voltage-fed and current-fed inverter drives. Scalar control is somewhat simple to implement, but the inherent coupling effect (i.e., both torque and flux are functions of voltage or current and frequency) gives sluggish response and the system is easily prone to instability because of a high-order (fifth-order) system effect. To make it more clear, if, for example, the torque is increased by incrementing the slip (i.e., the frequency), the flux tends to decrease. Note that the flux variation is always sluggish. The flux decrease is then compensated by the sluggish flux control loop feeding in additional voltage. This temporary clipping of flux reduces the torque sensitivity with slip and lengthens the response time. This explanation is also valid for current-fed inverter drives.

The foregoing problems can be solved by vector or field-oriented control. The invention of vector control in the beginning of 1970s, and the demonstration that an induction motor can be controlled like a separately excited dc motor, brought a renaissance in the high-performance control of ac drives. Because of dc machine-like performance, vector control is also known as decoupling, orthogonal, or transvector control. Vector control is applicable to both induction and synchronous motor drives. Undoubtedly, vector control and the corresponding feedback signal processing, particularly for modern sensorless vector control, are complex and the use of powerful microcomputer or DSP is mandatory. It appears that eventually, vector control will oust scalar control and will be accepted as the industry-standard control for ac drives.

2.5 Sensorless Vector Control

Sensorless vector control of an induction motor drive essentially means vector control without any speed sensor. An incremental shaft-mounted speed encoder (usually an optical type) is required for close loop speed or position control in both vector- and scalar-controlled drives. A speed signal is also required in indirect vector control in the whole speed range, and in direct vector control for the low-speed range, including the zero speed start-up operation. A speed encoder is undesirable in a drive because it adds cost and reliability problems, besides the need for a shaft extension and mounting arrangement. It is possible to estimate the speed signal from machine terminal voltages and currents with the help of a DSP. However, the estimation is normally complex and heavily dependent on machine parameters. Although sensorless vector-controlled drives are commercially available at this time, the parameter variation problem, particularly near zero speed, imposes a challenge in the accuracy of speed estimation.

2.6 Direct Torque and Flux Control (DTC)

In the mid-1980s, an advanced scalar control technique, known as direct torque and flux control (DTFC or DTC) or direct self-control (DSC), was introduced for voltage-fed PWM inverter drives. This technique was claimed to have nearly comparable performance with vector- controlled drives. Recently, the scheme was introduced in commercial products by a major company and therefore created wide interest. The scheme, as the name indicates, is the direct control of the torque and stator flux of a drive by inverter voltage space vector selection through a lookup table. Before explaining the control principle, we will first develop a torque expression as a function of the stator and rotor fluxes.

2.7 Adaptive Control

A linear control system with invariant plant parameters can be designed easily with the classic design techniques, such as Nyquist and Bode plots. Ideally, a vector-controlled ac drive can be considered as linear, like a dc drive system. However, in industrial applications, the electrical and mechanical parameters of the drive hardly remain constant. Besides, there is a load torque disturbance effect. For example, the inertia of an electric vehicle or subway drive will vary with passenger load. In a robotic drive, on the other hand, the inertia will change, depending on the length of the arm and the load it carries. In a rolling mill drive, the drive load torque will change abruptly when a metal slab is introduced within the rolls. Fig.5-3 (omitted) shows a block diagram of a speed-controlled vector drive indicating the moment of inertia J and load torque T_L variation. For the fixed control parameters in G, an increase of the J parameter will reduce the loop gain, deteriorating the system's performance. Similarly, a sudden increase of load torque T_L or J will temporarily reduce the speed until it is compensated by the sluggish speed loop. The effect of the parameter variation can be compensated to some extent by a high-gain negative feedback loop. But, excessive gain may cause an under-damping or instability problem in extreme cases. The problems discussed above require adaptation of the controller G in real time, depending on the plant parameter variation and load torque disturbance, so that the system response is not affected. Adaptive control techniques can be generally classified as

① Self-tuning control.

② MRAC.

③ Sliding mode or variable structure control.

④ Expert system control.

⑤ Fuzzy control.

⑥ Neural control.

Specialized English Words

megawatt　百万兆瓦

power supply　供电电源

scalar control　标量控制

orthogonal　正交的；直角的

harmonic　谐波

power factor　功率因素

adaptive control　自适应控制

cage-type machine　鼠笼式电机

Induction motor drive　感应电机驱动器

vector/field-oriented control　矢量 / 磁场定向控制

direct torque and flux control　直接转矩和磁通控制

sensorless vector control　无传感器矢量控制

voltage-fed inverter　电压源型逆变器

Unit 3 Process Control

3.1 Control Systems

The basic strategy by which a control system operates is logical and natural. In fact, the same strategy is employed in living organisms to maintain temperature, fluid flow rate, and a host of other biological functions. This is natural process control.

The technology of artificial control was developed using a human as an integral pan of the control action. When we learned how to use machines, electronics, and computers to replace the human function, the term automatic control came into use.

3.2 Process Control Principles

In process control, the basic objective is to regulate the value of some quantity. To regulate means to maintain that quantity at some desired value regardless of external influences. The desired value is called the reference value or setpoint.

In this section, a specific system will be used to introduce terms and concepts employed to describe process control.

The Process

Fig.5-4 (omitted) shows the process to be used for this discussion. Liquid is flowing into a tank at some rate, Q_{in}, and out of the tank at some rate, Q_{out}. The liquid in the tank has some height or level, h. It is known that the output flow rate varies as the square root of the height $Q_{out}=K\sqrt{h}$, so the higher the level, the faster the liquid flows out. If the output flow rate is not exactly equal to the input flow rate, the level will drop, if $Q_{out}>Q_{in}$, or rise, if $Q_{out}<Q_{in}$.

This process has a property called self-regulation. This means that for some input flow rate, the liquid height will rise until it reaches a height for which the output flow rate matches the input flow rate. A self-regulating system does not provide regulation of a variable to any particular reference value. In this example, the liquid level will adopt some value for which input and output flow rates are the same, and there it will stay. But if the input flow rate changed, then the level would change also, so it is not regulated to a reference value.

Example 1.1

The tank in Fig. 5-4 (omitted) has a reationship between flow and level given by

$Q_{out}=K\sqrt{h}$ where h is in feet and K=0.222(gal/min)/$ft^{1/2}$. Suppose the input flow rate is 2 gal/min. At what value of h will the level stabilize from self-regulation?

Solution

The level will stabilize from self-regulation when $Q_{out}=Q_{in}$. Thus, we solve for h,

$$h=\left(\frac{Q_{out}}{K}\right)^2=\left(\frac{2\text{gal/min}}{0.222(\text{gal/min})/ft^{1/2}}\right)^2=3ft \quad (5\text{-}1)$$

Suppose we want to maintain the level at some particular value, H, in Fig. 5-5 (omitted), regardless of the input flow rate. Then something more than self-regulation is needed.

Human-Aided Control

Fig. 5-6 (omitted) shows a modification of the tank system to allow artificial regulation of the level by a human. To regulate the level so that it maintains the value H, it will be necessary to employ a sensor to measure the level, This has been provided via a "sight tube", S, as shown in Fig. 5-7 (omitted). The actual liquid level or height is called the controlled variable. In addition, a valve has been added so that the output flow rate can be changed by a human. The output flow rate is called the manipulated variable or controlling variable.

Now the height can be regulated apart from input flow rate using the following strategy: the human measures the height in the sight tube and compares the value to the setpoint. If the measured value is larger, the human opens the valve a little to let the flow out increase, and thus the level lowers toward the setpoint. If the measured value is smaller than the setpoint, the human closes the valve a little to decrease the flow out and allow the level to rise toward the setpoint.

By a succession of incremental opening and closing of the value, the human can bring the level to the setpoint value, H, and maintain it there by continuous monitoring of the sight tube and adjustment of the value. The height is regulated.

Automatic Control

To provide automatic control, the system is modified as shown in Fig. 5-7 (omitted) so that machines, electronics, or computers replace the operations of the human. An instrument called a sensor is added, which is able to measure the value of the level and convert it into a proportional signal, s. This signal is provided as input to a machine, an electronic circuit, or a computer called the controller. The controller performs the function of the human in evaluating the measurement and providing an output signal, u, to change the valve setting via an actuator connected to the valve by a mechanical linkage.

When automatic control is applied to systems like the one in Fig. 5-7 (omitted), which are designed to regulate the value of some variable to a setpoint, it is called process control.

3.3 Servomechanisms

Another commonly used type of control system, which has a slightly different objective from process control, is called a servomechanism. In this case, the objective is to force some parameter to vary in a specific manner. This may be called a tracking control system. Instead of regulating a variable value to a setpoint, the servomechanism forces the controlled variable value to follow variation of the reference value.

For example, in an industrial robot arm like the one shown in Fig. 5-8 (omitted), servomechanisms force the robot arm to follow a path from point *A* to point *B*. This is done by controlling the speed of motors driving the arm and the angles of the arm parts.

The strategy for servomechanisms is similar to that for process-control systems, but the dynamic differences between regulation and tracking result in differences in design and operation of the control system.

3.4 Discrete-State Control Systems

This is a type of control system connected with controlling a sequence of events rather than regulation or variation of individual variables. For example, the manufacture of paint might involve the regulation of many variables, such as mixing temperature, flow rate of liquids into mixing tanks, speed of mixing, and so on. Eath of these might be expected to be regulated by process-control loops. But there is also a sequence of event that must occur in the overall process of manufacturing the paint. This sequence is described in terms of events that are aimed to be started and stopped on a specified schedule. Referring to the paint example, the mixture needs to be heated with a regulated temperature for a certain length of time and then perhaps pumped into a different tank and stirred for another period.

The starting and stopping of events is a discrete-base system because the event is either true or false, i.e., started or stopped, open or closed, on or off. This type of control system can also be made automatic and is perfectly suited to computer based controllers.

These discrete-state control systems are often implemented using specialized computer based equipment called programmable logic controllers (PLCs).

Specialized English Words

process control 过程控制	logical 逻辑的
biological function 生物学功能	artificial control 人工控制
self-regulation 自动调节	human-aided control 半自动控制

manipulated variable　操纵变量　　servomechanism　伺服机构；自动驾驶装置
setpoint　设定点　　reference value　参考值
programmable logic controller (PLC)　可编程逻辑控制器

Unit 4　Computer Control Technology

4.1 Introduction to Digital Control

As a result of the revolution in the cost-effectiveness of digital computers, there has been an increasing use of digital logic in embedded applications such as controllers in feedback systems. A digital controller gives the designer much more flexibility to make modifications to the control law after the hardware design is fixed because the formula for calculating the control signal is in the software rather than the hardware. In many instances, this means that the hardware and software designs can proceed almost independently, saving a great deal of time. Also, it is relatively easy to include binary logic and nonlinear operations as part of the function of a digital controller as compared to an analog controller. Special processors designed for real-time signal processing and known as digital signal processors (DSPs) are particularly well suited for use as real-time controllers. Here, we give a brief introduction to the most simple techniques for digital designs.

A digital controller differs from an analog controller in that the signals must be sampled and quantized. A signal to be used in digital logic needs to be sampled first and then the samples need to be converted by an analog-to-digital converter or A/D into a quantized digital number. Once the digital computer has calculated the proper next control signal value, this value needs to be converted back into a voltage and held constant or otherwise extrapolated by a digital-to-analog converter or D/A in order to be applied to the actuator of the process. The control signal is not changed until the next sampling period. As a result of sampling, there are strict limits on the speed and bandwidth of a digital controller. A reasonable rule of thumb for selecting the sampling period is that during the rise time of the response to a step, the input to the discrete controller should be sampled approximately six times. By adjusting the controller for the effects of sampling, the sample period can be as large as two to three times per rise time. This corresponds to a sampling frequency that is 10 to 20 times the system's closed-loop bandwidth. The quantization of the controller signals introduces an equivalent extra noise into the system and to keep this interference at an acceptable level, the A/D converter usually has an accuracy of 10 to 12 bits although inexpensive systems have been designed with only 8 bits. For a first analysis, the effects of the quantization are usually ignored, as they will be in this introduction. A simplified block diagram of a system with a digital controller is shown in Fig. 5-9 (omitted).

For this introduction to digital control, we will describe a simplified technique for finding a discrete (sampled but not quantized) equivalent to a given continuous controller. The method depends on the sampling period, T_s, being short enough that the reconstructed control signal is close to the signal that the original analog controller would have produced. We also assume that the numbers used in the digital logic have enough accurate bits so that the quantization implied in the A/D and D/A processes can be ignored. While there are good analysis tools to determine how well these requirements are met, here we will test our results by simulation, following the well-known advice that "The proof of the pudding is in the eating."

Finding a discrete equivalent to a given analog controller is equivalent to finding a recurrence equation for the samples of the control, which will approximate the differential equation of the controller. The assumption is that we have the transfer function of an analog controller and wish to replace it with a discrete controller that will accept samples of the controller input, $e(kT_s)$, from a sampler and, using past values of the control signal, $u(kT_s)$ and present and past samples of the input, $e(kT_s)$ will compute the next control signal to be sent to the actuator. As an example, consider a PID controller with the transfer function

$$U(s)=\left(k_P+\frac{k_I}{s}+k_D s\right)E(s) \tag{5-2}$$

which is equivalent to the three terms of the time-domain expression

$$u(t)=k_P e(t)+k_I\int_0^t e(\tau)\mathrm{d}\tau+k_D\dot{e}(t) \tag{5-3}$$

$$=u_P+u_I+u_D \tag{5-4}$$

Based on these terms and the fact that the system is linear, the next control sample can be computed term-by-term.

The proportional term is immediate:

$$u_P(kT_s+T_s)=k_P e(kT_s+T_s) \tag{5-5}$$

The integral term can be computed by breaking the integral into two parts and approximating the second part, which is the integral over one sample period, as follows.

$$u_I(kT_s+T_s)=k_I\int_0^{kT_s+T_s} e(\tau)\mathrm{d}\tau \tag{5-6}$$

$$=k_I\int_0^{kT_s} e(\tau)\mathrm{d}\tau+k_I\int_{kT_s}^{kT_s+T_s} e(\tau)\mathrm{d}\tau \tag{5-7}$$

$$= u_I(kT_s) + \{area\ under\ e(\tau)\ over\ one\ period\} \tag{5-8}$$

$$\cong u_I(kT_s) + k_I \frac{T_s}{2}\{e(kT_s + T_s) + e(kT_s)\} \tag{5-9}$$

In Eq. (5-9) the area in question has been approximated by that of the trapezoid formed by the base T_s and vertices $e(kT_s + T_s)$ and $e(kT_s)$ as shown by the dashed line in Fig. 5-10 (omitted).

The area can also be approximated by the rectangle of amplitude $e(kT_s)$ and width T_s shown by the solid blue in Fig. 5-10 (omitted) to give $u_I(kT_s + T_s) = u_I(kT_s) + k_I T_s e(kT_s)$.

In the derivative term, the roles of u and e are reversed from integration and the consistent approximation can be written down at once from Eq. (5-8) and Eq. (5-9) as

$$\frac{T_s}{2}\{u_D(kT_s + T_s) + u_D(kT_s)\} = k_D\{e(kT_s + T_s) - e(kT_s)\} \tag{5-10}$$

As with linear analog transfer functions, these relations are greatly simplified and generalized by the use of transform ideas. At this time, the discrete transform will be introduced simply as a prediction operator z much as if we described the Laplace transform variable, s, as a differential operator. Here we define the operator z as the forward shift operator in the sense that if $U(z)$ is the transform of $u(kT_s)$ then $zU(z)$ will be the transform of $u(kT_s + T_s)$. With this definition, the integral term can be written as

$$zU_I(z) = U_I(z) + k_I \frac{T_s}{2}[zE(z) + E(z)] \tag{5-11}$$

$$U_I(z) = k_I \frac{T_s}{2}\frac{z+1}{z-1}E(z) \tag{5-12}$$

and from Eq. (5-12), the derivative term becomes the inverse as

$$U_D(z) = k_D \frac{2}{T_s}\frac{z-1}{z+1}E(z) \tag{5-13}$$

The complete discrete PID controller is thus described by

$$U(z) = \left(k_P + k_I \frac{T_s}{2}\frac{z+1}{z-1} + k_D \frac{2}{T_s}\frac{z-1}{z+1}\right)E(z) \tag{5-14}$$

Specialized English Words

computer control 计算机控制　　digital computer 数字计算机

analog controller 模拟控制器　　real-time 实时

signal processing 信号处理	digital signal processor (DSP) 数字信号处理器
bandwidth 带宽	rise time 上升时间
sampling period 采样时间	transfer function 传递函数
integral term 积分项	analog-to-digital converter 模/数转换器

Unit 5 Industrial Robots

Industrial robots are considered as a cornerstone of competitive manufacturing, which aims to combine high productivity, quality, and adaptability at minimal cost. In 2007, more than one million industrial robot installations were reported, with automotive industries as the predominant users with a share of more than 60%. However, high-growth industries (in life sciences, electronics, solar cells, food, and logistics) and emerging manufacturing processes (gluing, coating, laser-based processes, precision assembly, etc.) will increasingly depend on advanced robot technology. These industries' share of the number of robot installations has been growing steadily.

The production of industrial robots on the one hand, and the planning, integration, and operation of robot workcells on the other hand, are largely independent engineering tasks. In order to be produced in sufficiently large quantities, a robot design should meet the requirements for the widest set of potential applications. As this is difficult to achieve in practice, various classes of robot designs regarding payload capacity, number of robot axes, and workspace volume have emerged for application categories such as assembly, palletizing, painting, welding, machining, and general handling tasks.

Generally, a robot workcell consists of one or more robots with controllers and robot peripherals: grippers or tools, safety devices, sensors, and material transfer components for moving and presenting parts. Typically, the cost of a complete robot workcell is four times the cost of the robots alone.

A robot workcell is usually the result of customized planning, integration, programming, and configuration, requiring significant engineering expertise. Standardized engineering methods, tools, and best-practice examples have become available to reduce costs and provide more predictable performance.

Today's industrial robots are mainly the result of the requirements of capital-intensive large-volume manufacturing, mainly defined by the automotive, electronics, and electrical goods industries. Future industrial robots will not be a mere extrapolation of today's designs with respect to features and performance data, but will rather follow new design principles addressing a wider range of application areas and industries. At the same time, new technologies, particularly from the information technology (IT) world, will have an increasing impact on the design, performance, and cost of future industrial robots.

International and national standards now help to quantify robot performance and define safety precautions, geometry, and media interfaces. Most robots operate behind secure barriers to keep people at a safe distance. Recently, improved safety standards have allowed direct human-robot collaboration, permitting robots and human factory workers to share the same workspace in safety.

We will first present the basic principles that are used in industrial robotics and a review of programming methods. We will also discuss tools (end-effectors) and system integration requirements. The unit will be closed with the presentation of selected, unsolved problems that currently inhibit the wider application of industrial robots.

5.1 Kinematics and Mechanisms

The choice of mechanism, its kinematic properties, the computation methods used to determine joint motions, and the intended application of a robot manipulator are all closely related. The diagrams in Fig. 5-11 (omitted) show several common types of robot mechanism. With advances in the state of the art in kinematic algorithms and computer hardware processing capabilities, computation is much less of a constraint on mechanism choice than it was for early robot designers. The choice of mechanical structure of the robot depends mostly on fundamental mechanical requirements such as payload and workspace size. Considering a given level of cost, there is usually a tradeoff between workspace size and stiffness. To enable the robot to reach inside or around obstacles it is clearly advantageous to use an articulated mechanical design.

Considering also the stiffness and accuracy (in a practical sense considering what is reasonable to build), the picture is more complex. Each of the first three types in Fig. 5-12 (omitted) we refer to as serial kinematic machines (SKMs), while the last is a parallel kinematic machine (PKM). To obtain maximum stiffness, again for a certain minimum level of cost, the end-effector is better supported from different directions, and here the PKM has significant advantages. On the other hand, if high stiffness (but not low weight and high dexterity) is the main concern, a typical computerized numerical control (CNC) machine (e.g., for milling) is identical in principle to the gantry mechanism. There are also modular systems with servo-controlled actuators that can be used to build both robots with purpose-designed mechanisms.

5.2 End-Effectors and System Integration

It is interesting to note that connecting different workcell devices with each other, and integrating them into a working system, is hardly mentioned in the robotics literature. Nevertheless, in actual nontrivial installations, this part typically represents half of the cost of the installation. The automation scenario includes all the problems of integrating computers and their peripherals, plus additional issues that have to do with the variety of (electrically and mechanically incompatible) devices and their interaction with the physical environment (including its inaccuracies, tolerances, and unmodeled physical effects such as backlash and friction). The number of variations is enormous so it is often not possible to create reusable solutions. In total this results in a need for extensive engineering to put a robot to work. This engineering is what we call system integration.

Carrying out system integration according to current practice is not a scientific problem as such (although how to improve the situation is), but the obstacles it comprises form a barrier to applying advanced sensor-based control for improved flexibility, as is needed in short-series production. In particular, in future types of applications using external sensing and high-performance feedback control within the workcell, system integration will be an even bigger problem since it includes tuning of the feedback too.

For long production runs, the engineering cost of system integration is less of a problem since its cost per manufactured part is small. On the other hand, the trend towards shorter series of customized products, or products with many variants that are not kept in stock, calls for high flexibility and short changeover times. Flexibility in this context refers to variable product variants, batch sizes, and process parameters. In particular this is a problem in small and medium-sized productions, but the trend is similar for larger enterprises as flexibility requirements are continuously on the rise. One may think that simply by using well standards for input/output (I/O) and well-defined interfaces, integration should be just a matter of connecting things together and running the system. Let us examine why this is not the case.

Example 5.1 Step 1 of 3:

Simple pick-and-insert (assembly) operation. As a first step let us consider only one robot performing pick-and-place from/to known locations, and assume it is a well-known object such that we can use one specific type of gripper (such as a gripper, in this case based on a vacuum cup). We then only need one digital output from the robot controller. Let us call this output GRIPPER (i.e., the name of the number of the output) and the values GRASP and RELEASE for the high and low signals depending on the sign of the hardware connection. An example of a robot program can be found in the previous section.

Of course the object-oriented software solution would be to have a gripper class with operations grasp and release. That would be appropriate for a robot simulator in pure software, but for integration of real systems such encapsulation of data (abstract data types) would introduce practical difficulties because:

(1) values are explicit and accomplished by external (in this case) hardware, and for testing and debugging we need to access and measure them;

(2) online operator interfaces permit direct manipulation of values, including reading of output values. The use of functions of object methods (so-called set: ers and get:ers) would only complicate the picture; maintaining consistency with the external devices (such as the definitions of the constants GRIPPER, GRASP, and RELEASE) would be no simpler.

Already in this toy example we can see that the user I/O configuration depends on both connected external devices (the gripper), and also on what I/O modules (typically units on some kind of field bus) that are installed in the robot controller (which may be subject to change when another task is given to the robot). Returning to our example, selecting modules and interfaces extends to the items of Tab. 5-1 (omitted), where the application and task levels relate to robot programming and the research issue of how to simplify system integration, respectively.

Example 5.1 Step 2 of 3:

Simple pick-and-insert (assembly) operation. Assume we want to use a probe to detect the type and the location of the workpiece, that we need several different types of grippers depending on what object we should handle, and that the place operation needs to be a force-controlled insertion. Additionally, since we need different tools for different operations, we need a tool exchanger.

With these additional requirements we need more mechanical interfaces, the load/weight/performance considerations are more difficult (the weight of the tool exchanger and adaptor plate decreases the net payload for the grasped objects), and several more electrical interfaces need to be included. Equipment examples are shown in Fig. 5-13 (omitted), an example that extends the pick-and-insert case to include unknown positions and a gripper that can grab several types of pieces (including a wrist force sensor and force feedback in the robot fingers); details concerning robot programming and device configurations are omitted for brevity. In the presented example a three-finger robotic hand is used to illustrate that these advanced and programmable tools are finding their way into industrial applications, for example, the hand used here has position and force-sensing capabilities. When grasping an object, the finger base and tip links move together. If a base link encounters an object it stops, while the tip link keeps moving until it makes contact with the object as well. If the tip links encounter an object first, the whole finger assembly stops moving. The example depicted in Fig. 5-14 (omitted) shows a setup where the robot can pick the workpiece from an unknown position; the working pieces are identified using a CCD camera that returns the number of pieces and their position. The robot is then commanded to pick one and update its information from the cell. This example uses two tasks: one to receive remote commands, and the other to implement the gripper and camera services.

Example 5.1 Step 3 of 3:

Simple pick-and-insert (assembly) operation. As a final requirement, to monitor the quality of the assembly operation, assume that we want to have slip detection embedded into the gripper and that we need logging of the force control signal. The production statistics should be provided on the overall plant level of the factory.

These additional requirements call for integration of low-level devices with high-level factory control system, and is hence called vertical integration, whereas integration within (for instance) the workcell level is called horizontal integration. Lack of self-descriptive and self-contained data descriptions that are also useful at the real-time level further increases the integration effort since data interfaces/conversions typically have to be manually written. In some cases, as illustrated in Fig. 5-15 (omitted), there are powerful software tools available for the integration of the robot user level and the engineering level. The fully (on all levels) integrated and nonproprietary system, sometimes referred to as a digital factory, still is a challenge, particularly for small and medium-sized productions.

5.3 Conclusions and Long-Term Challenges

The widespread use of robots in standard, large scale production such as the automotive industry, where robots (even with impressive performance, quality, and semiautomatic programming) perform repetitive tasks in very well-known environments, for some time resulted in the common opinion that industrial robotics is a solved problem. However, these applications comprise only a minor part of the industrial work needed in any wealthy society, especially considering the number of companies and the variety of applications. The use of robots in small and medium-sized manufacturing is still tiny.

Global prosperity and wealth requires resource-efficient and human-assistive robots. The challenges today are to recognize and overcome the barriers that are currently preventing robots from being more widely used.

Taking a closer look at the scientific and technological barriers, we find the following challenges:

(1) Human-friendly task specification, including intuitive ways of expressing permitted/normal/expected variations. That is, there are many upcoming and promising techniques for user-friendly human robot interaction (such as speech, gestures, manual guidance, and so on), but the focus is still on specification of the nominal task. The foreseen variations, and the unforeseen variations experienced during robot work, are more difficult to manage. When instructing a human he/she has an extensive and typically implicit knowledge about the work and the involved processes. To teach a robot, it is an issue of both how to realize what the robot does not know, and how to convey the missing information efficiently.

(2) Efficient mobile manipulation. Successful implementations and systems are available for both mobility and for manipulation, but accomplished in different systems and using different types of (typically incompatible) platforms. A first step would be to accomplish mobile manipulation at all, including the combination of legged locomotion (for stairs and rough terrain), autonomous navigation (with adaptive but predictable understanding of constraints), dexterous manipulation, and robust force/torque interaction with environments (that have unknown stiffness). As a second step, all this needs to be done with decent performance using reasonably priced hardware, and with interfaces according to the previous item. Thus, we are far from useful mobile manipulation.

(3) Low-cost components including low-cost actuation. Actuation of high performance robots represent about a third of the overall robot cost, and improved modularity often results in a higher total hardware cost (due to less opportunities for mechatronic optimization). On the other hand, cost-optimized (with respect to certain applications) systems result in more-specialized components and smaller volumes, with higher costs for short-series production of those components. Since future robotics and automation solutions might provide the needed cost efficiency for short-series customized components, we can interpret this as a boot-strapping problem, involving both technical and business aspects. The starting point is probably new core components that can fit into many types of systems and applications, calling for more mechatronics research and synergies with other products.

(4) Composition of subsystems. In most successful fields of engineering, the principle of superposition holds, meaning that problems can be divided into subproblems and that the solutions can then be superimposed (added/combined) onto each other such that the total solution comprises a solution to the overall problem. These principles are of key importance in physics and mathematics, and within engineering some examples are solid-state mechanics, thermal dynamics, civil engineering, and electronics. However, there is no such thing for software, and therefore not for mechatronics (which includes software) or robotics (programmable mechatronics) either. Thus, composition of unencapsulated subsystems is costly in terms of engineering effort. Even worse, the same applies to encapsulated software modules and sub-

systems. For efficiency, system interconnections should go directly to known (and hopefully standardized) interfaces, to avoid the indirections and extra load (weight, maintenance, etc.) of intermediate adapters (applying to both mechanics for end-effector mounting and to software). Interfaces can be agreed upon, but the development of new versions typically maintains backward compatibility (newer devices can be connected to old controllers), while including the reuse of devices calls for mechanisms for forward compatibility (automatic upgrade based on meta information of new interfaces) to cover the case that a device is connected to a robot that is not equipped with all the legacy or vendor-specific code.

(5) Embodiment of engineering and research results. Use or deployment of new technical solutions today still starts from scratch, including analysis, understanding, implementation, testing, and so on. This is the same as for many other technical areas, but the exceptional wide variety of technologies involved with robotics and the need for flexibility and upgrading makes it especially important in this field. Embodiment into components is one approach, but knowledge can be applicable to engineering, deployment, and operation, so the representation and the principle of usage are two important issues. Improved methods are less useful if they are overly domain specific or if engineering is experienced to be significantly more complicated. Software is imperative, as well as platform and context dependent, while know-how is more declarative and symbolic. Thus, there is still a long way to go for efficient robotics engineering and reuse of know-how.

(6) Open dependable systems. Systems need to be open to permit extensions by third parties, since there is no way for system providers to foresee all upcoming needs in a variety of new application areas. On the other hand, systems need to be closed such that the correctness of certain functions can be ensured. Extensive modularization in terms of hardware and supervisory software makes systems more expensive and less flexible (contrary to the needs of openness). Highly restrictive frameworks and means of programming will not be accepted for widespread use within short-time-to-market development. Most software modules do not come with formal specification, and there is less understanding of such needs. Thus, systems engineering is a key problem.

(7) Sustainable manufacturing. Manufacturing is about transformation of resources into products, and productivity (low cost and high performance) is a must. For long-term sustainability, however, those resources in terms of materials and the like must be recycled. In most cases this can be achieved by crushing the product and sorting the materials, but in some cases disassembly and automatic sorting of specific parts are needed. There is therefore a need for robots in recycling and demanufacturing. Based on future solutions to the above items, this is then a robot application challenge.

An overall issue is how both industry and academia can combine their efforts such that sound business can be combined with scientific research so that future development overcomes the barriers that are formed by the above challenges.

Specialized English Words

industrial robot　工业机器人
solar cell　太阳能电池
workcell　工作单元
sensor　传感器
end-effector　末端执行器
system integration　系统集成
serial kinematic machine（SKM）　串行运动机器
parallel kinematic machine（PKM）　并行运动机器
computerized numerical control（CNC）　电脑数控

Unit 6 科研交流的演说技巧

6.1 学术演讲的技巧和要领

学术演讲是学术会议、学术汇报、学术研讨会、学术讲座等场合常见的一种交流形式，是科研工作者有目的地向与会者进行学术信息传播和交流。一场成功的专业学术演讲并不是机械、生硬地把学术报告的内容转移到幻灯片，而是包括信息转换、媒体转换、环境营造等一系列过程。学术演讲的目的是在真正意义上对于某个主题或某项研究进行交流。为了达到这个目的，需要注意以下几点。

第一，分析并了解观众。演讲是否成功有效，在于观众所吸收的东西。因此，需要分析和了解观众的兴趣、专业背景和期待，从观众角度设计演讲、调整演讲的内容，使它容易被理解和接受、生动有趣，以此来吸引观众的注意力和兴趣，从而实现演讲者和观众的双向互动。

第二，组织演讲内容。明确演讲目的和选定信息要点。学术演讲的目的通常是将一个科技问题以及创新解决该问题的思路、过程讲清楚。要定义好问题的框架；设计有效的开场白，例如，用统计资料、故事或实际经验引入问题；合理安排演讲正文的表达结构和论证方式，例如总分总或总分的结构、递进式或并列式的论证方式；演讲结尾可以是归纳要点、首尾呼应或者陈述结论。

第三，呈现演讲内容。成功、有效的学术演讲要求演讲者必须将学术演讲的原信息转换成观众容易理解和接受的文本形态。把报告的问题或成果解析化、具体化和形象化成一个个小问题或小成果。采用提纲、表格、框架图、箭头图、简图，甚至图片、视频等图式化、视频化方式，把这些小问题或小成果按逻辑顺序和思维方式进行归纳，以条理化、精炼化、图表化的文图搭配简单文稿，再以口述和多媒体演示相结合的方式呈现出来。深入浅出地以通俗、简明的形态解释高深的学术理论、表达繁杂的程序或思想。还可以采用视听化的演讲方法，这样能准确、快速、大量地传递信息给听众。总之，就是少说多演。

第四，其他技巧。着装得体，展现希望传递给观众的形象，整体搭配符合演讲的文化背景；合理的面部表情管理，表情自然而生动，并且能符合演讲当时的情形；自信和真诚的目光交流，尽量与每位观众都有目光接触，保持 2 ～ 3 秒，再慢慢从一个人移动到另外一个人身上；演讲时姿势轻松、自然，不要左右摇摆、前后晃动；利用手势辅助演讲内容，把演讲的内容比画成直观的现象，或利用手势强调正在演讲的内容，注意手的小动作。

6.2 英文演讲幻灯片制作

第一，合理规划每张幻灯片的时间。根据演讲规定的时间安排演讲内容、幻灯片的张数和时间，每张幻灯片展示尽量少而足够表达观点的信息，一般来说，每张幻灯片最多一分钟，但是具体的时间应该由幻灯片内容来决定。注意不要快速掠过幻灯片，给观众留下不好的印象。

第二，设定每张幻灯片的目标。 由于演讲时间和演讲内容有限，所以必须认真、清晰地设定每张幻灯片的目标和要点。确保每张幻灯片的目标和要点之间都有逻辑顺序，并且这些要点可以完整、有条理地支撑整个演讲内容。

第三，设定演讲提纲页。一般地，在演讲幻灯片的标题页后会紧跟一个带有纲要或目录的幻灯片，可以帮助观众在短时间内迅速明确和记住演讲的每个主题。若演讲时间或内容较简短，则不需要提纲页。

第四，避免幻灯片内容过多。幻灯片是演讲的一个可视化的多媒体辅助，而不是演讲的主要部分。一般来说，每一张幻灯片都要有标题，标题不宜过长。幻灯片内容不要用大段的文字，尽量避免完整的句子，可以采用标题列表或示意图等方式代替大段文字。注意，图表中需要标注。

第五，避免幻灯片设计过于复杂。根据演讲内容选择合适的幻灯片背景和配色方案，否则会弱化幻灯片的内容。一张幻灯片中，动画等多媒体不宜过多，以免影响或分散观众的注意力。注意合理安排多媒体的播放时间。

第六，合理安排幻灯片的文字。 幻灯片上的书面文字要包含演讲的关键字，演讲的内容和顺序要尽可能地匹配幻灯片上的书面文字。有时可以借助箭头指向文字，配合演讲，从而一步步引导观众理解演讲内容。

注意事项：

（1）模板的风格要注意严肃性、新颖性，可以采用深底浅字。

（2）模板的基本要求：尽量选一个底色统一的模板；整个幻灯片配色一致，字体、行距一致；不要过多地使用动画。

（3）幻灯片的文字：文字不能太多；每张幻灯片应有标题和正文。

（4）研究过程往往需要用幻灯片的流程图表达，但插入剪贴画勿太多，以免减少学术气氛。

（5）复杂的化学式和数学公式如果有的话最好写在幻灯片上，这样演讲者就不用把它读出来了。

（6）注意演讲语调、语速和音量。

6.3 学术会议问答讨论环节的口语

一般来说，学术演讲之后有一个问答讨论环节，该环节是观众针对演讲内容进行具体的提问和探讨，通过现场的信息反馈有时可以得到建设性建议，对下一步的科研工作有很大的促进作用。

1. 答问的方式与技巧

由于时间的关系，演讲者在现场往往没有充分的时间考虑问题。因此，在学术演讲前，演讲者可以预测观众可能提出的问题，并准备好合适的解答和幻灯片说明。演讲者应合理控制整个讨论的过程，把握每位提问者的提问时间，通过目光接触获得提问者和观众的反馈，确保给予提问者和观众满意的回答。礼貌地回避与主题无关的问题。回答问题之前可以适当地重复、确认问题，确保对问题的正确理解，增加思考的时间。回答问题要切题，运用事实、数据等客观地证明自己的观点，尽可能多地提及报告中的内容，这样有利于加深观众对演讲内容的印象。对于过于复杂的深层问题无法简单或完整地回答时，可以在简略回答之后建议提问者选择其他时间或其他方式与自己讨论。

2. 常用表达

① 提问

I would like to ask (sb.) a question about…

I would like to address/raise a question about… (to sb.) .

I wonder if you would be kind/good enough to explain/comment on/elaborate something about …

I am curious about/interested in/anxious to…

Could you please tell me whether/why/how/what/when…

Would you please give me/us an example/more information about…

Do you see any difference/relation between ... and ...

② 回答

Are you asking me the question about…

I am sorry. I am not quite clear about the question? Do you mean…

Could/Would you address your question more specifically?

Thank you for asking the question.

Exactly!/Precisely!/Certainly!/Surely!

Let me answer your question by using ... as an example.

Let me answer your question by making a few comments on/explanation about…

实训课程五：科研报告的幻灯片制作及其演说

下面以一份英文演讲幻灯片为例（摘自学生作业），介绍科研报告的幻灯片制作及其演说。

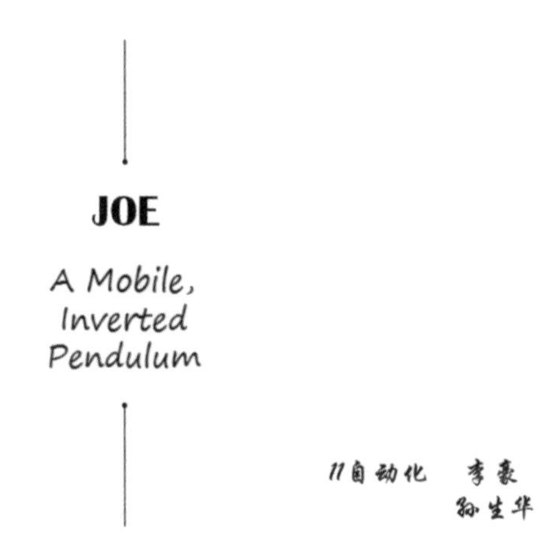

1. 演讲开场白

问候听众，介绍自己，并告知演讲主题。在英语演讲的开头问候听众是最基本的礼貌。在一个陌生的场合做英语演讲时，英语演讲稿的开头可以先介绍自己。英语演讲的开头要记得告知听众演讲的主题。一开始就告诉听众主题，让他们带着已有的自我认知跟着你的演讲往下听，他们的兴趣会更浓厚。

如：Good morning/afternoon/evening ladies and gentleman, Thank you for being on time/making the effort to come today. Let me introduce myself first. My name is ... from ... Thank you for giving me the opportunity to tell you about ... The subject of my presentation is ...

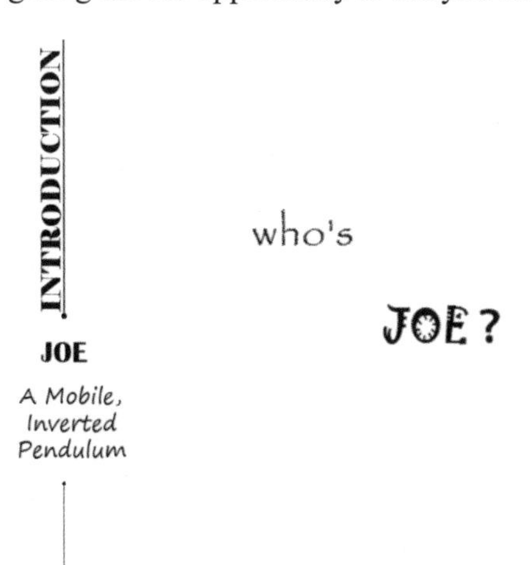

2. 引入

在英语演讲稿的开头引入一些能引起听众兴趣的内容是演讲技巧之一。

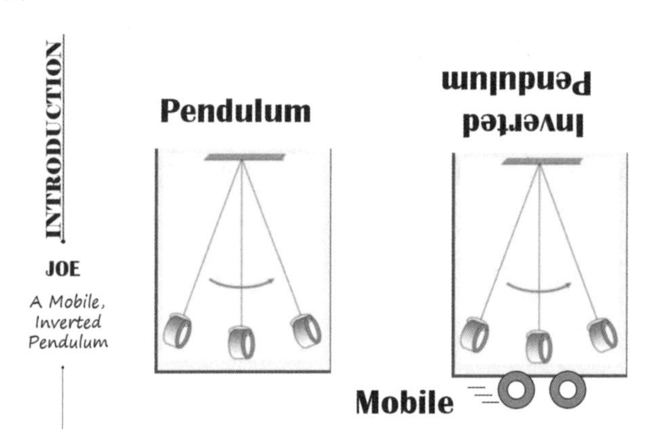

3. 简介

在开始的一分钟里要清楚说明论文的研究领域和研究问题；将论文中的问题用通俗易懂的语言与一些日常生活联系起来，以引起听众的兴趣，使之了解研究结果的价值；不常用或重要的特殊词语要在使用之前加以说明；比较难的专业术语要用简单的定义、类比等方法或图片来说明。

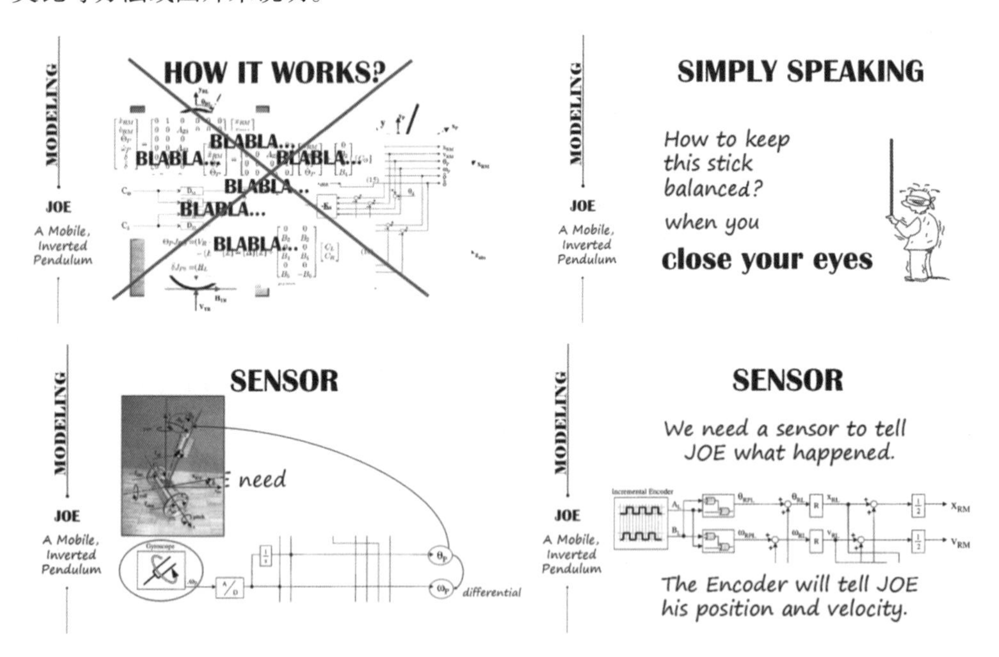

4. 主体部份

主体部分不能有过多的内容，否则要全部说清楚，时间会很匆忙；不要只是简单陈述结果，而要阐述它们的意义是好还是坏、有多大价值、有没有惊人发现；演讲者不能陷入讲解枯燥细节的泥潭中，如果要介绍细节，要把握好节奏。细节的介绍仅仅是为了便于专家理解研究的关键点；复杂的化学式和数学公式如果有的话要在幻灯片上写出来，在谈到时口头上以其名或“这个公式”称之，并指着幻灯片来说明，这样可以节省时间。

5. 结论

演讲者在结论部分的演讲正式开始前要用“conclusion”（结论）、“summary”（总结）、“ending”（结束语）或其他简短的词语来提醒听众们演讲将要结束；结论部分要用一句话来总结主体部分解释过的2～3个要点；演讲者要突出强调论文中最重要的研究结论；演讲者要给出研究结果的实用性或者说出研究的必要性和意义；结论部分至少要以一个开放式的问题或研究方向结束，以作为将来的研究方向。

Chapter 6

SMART GRID

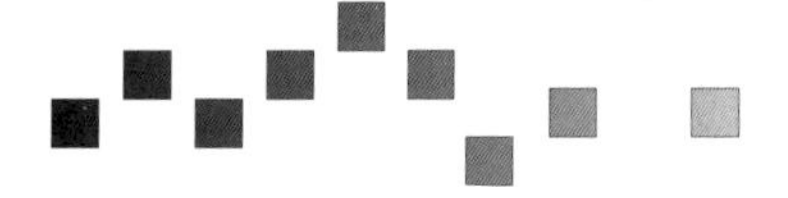

Unit 1 Signal and System

1.1 What Is a Signal?

Signals, in one form or another, constitute a basic ingredient of our daily lives. For example, a common form of human communication takes place through the use of speech signals, in a face-to-face conversation or over a telephone channel. Another common form of human communication is visual in nature, with the signals taking the form of images of people or objects around us.

Yet another form of human communication is electronic mail over the Internet. In addition to providing mail, the Internet serves as a powerful medium for searching for information of general interest, for advertising, for telecommuting, for education, and for playing games. All of these forms or communication over the Internet involve the use of information-bearing signals of one kind or another. Other real life examples in which signals of interest arise are discussed subsequently.

By listening to the heartbeat of a patient and monitoring his or her blood pressure and temperature, a doctor is able to diagnose the presence or absence of an illness or disease. The patient's heartbeat and blood pressure represent signals that convey information to the doctor about the state of health of the patient.

In listening to a weather forecast over the radio, we hear references made to daily variations in temperature, humidity, and the speed and direction of prevailing winds. The signals represented by these quantities help us, for example, to form an opinion about whether to stay indoors or go out for a walk.

The daily fluctuations in the prices of stocks and commodities on world markets, in their own ways, represent signals that convey information on how the shares in a particular company or corporation are doing. On the basis of this information, decisions are made regarding whether to venture into new investments or sell off old ones.

A probe exploring outer space sends valuable information about a faraway planet back to a station on Earth. The information may take the form of radar images representing surface profiles of the planet, infrared images conveying information on how hot the planet is, or optical images revealing the presence of clouds around the planet. By studying these images, our knowledge of the unique characteristics of the planet in question is enhanced significantly.

A signal is formally defined as function of one or more variables that convey information on the nature of a physical phenomenon. When the function depends on a single variable, the signal is said to be one dimensional. A speech signal is an example of a one-dimensional signal whose amplitude varies with time, depending on the spoken word and who speaks it. When the function depends on two or more variables, the signal is said to be multidimensional. An image is an example of a two dimensional signal, with the horizontal and vertical coordinates of the image representing the two dimensions.

1.2 What Is a System?

In the example of signals mentioned in the preceding section, there is always a system associated with the generation of each signal and another system associated with the extraction of information from the signal. For example, in speech communication, a sound or signal excites the vocal tract, which represents a system. The processing of speech signals usually relies on the use of our ears and auditory pathways in the brain. In this case, the systems responsible for the production and reception of signals are biological in nature. These systems could also be implemented using electronic systems that try to emulate or mimic their biological counterparts. For example, the processing of a speech signal may be performed by an automatic speech recognition system in the form of a computer program that recognizes words or phrases.

A system does not have a unique purpose. Rather purpose depends on the application of interest. In an automatic speaker recognition system, the function of the system is to extract information from an incoming speech signal for the purpose of recognizing or identifying the speaker. In a communication system, the function of the system is to transport the information contained in a message over a communication channel and deliver that information to a destination in a reliable fashion. In an aircraft landing system, the requirement is to keep the aircraft on the extended centerline of a runway.

A system is formally defined as an entity that manipulates one or more signals to accomplish a function, thereby yielding new signals. The interaction between a system and its associated signals is illustrated schematically in Fig. 6-1 (omitted). Naturally, the descriptions of the input and output signals depend on the intended application of the system:

(1) in an automatic speaker recognition system, the input signal is a speech (voice) signal, the system is a computer, and the output signal is the identity of the speaker.

(2) in a communication system, the input signal could be a speech signal or computer data, the system itself is made up of the combination of a transmitter, channel, and receiver, and the output signal is an estimate of the information contained in the original message.

(3) in an aircraft landing system, the input signal is the desired position of the aircraft relative to the runway, the system is the aircraft, and the output signal is a correction to the lateral position of the aircraft.

Specialized English Words

signal and system　信号与系统
humidity　湿度
planet　行星
optical image　光学图像
vocal tract　声道
channel　通道
telecommute　远距离工作
radar　雷达
infrared image　红外图像
horizontal/vertical coordinate　水平/垂直轴
transmitter　发射机
receiver　接收机

Unit 2　Communication Technology of Power System

2.1 Introduction

The communication infrastructure of a power system typically consists of SCADA (supervisory control and data acquisition) systems with dedicated communication channels to and from the System Control Centre and a Wide Area Network (WAN). Some long-established power utilities may have private telephone networks and other legacy communication systems. The SCADA systems connect all the major power system operational facilities, that is, the central generating stations, the transmission grid substations and the primary distribution substations to the System Control Centre. The WAN is used for corporate business and market operations. These form the core communication networks of the traditional power system. However, in the Smart Grid, it is expected that these two elements of communication infrastructure will merge into a Utility WAN.

An essential development of the Smart Grid (Fig. 6-2 (omitted)) is to extend communication throughout the distribution system and to establish two-way communications with customers through Neighbourhood Area Networks (NANs) covering the areas served by distribution substations. Customers' premises will have Home Area Networks (HANs). The interface of the Home and Neighbourhood Area Networks will be connected through a smart meter or smart interfacing device.

In the ISO/OSI reference model, the upper layers deal with Applications of the data irrespective of its actual delivery mechanism while the lower layers look after delivery of information irrespective of its Application. In this unit, communication technologies that are associated with the lower three layers of the ISO/OSI reference model are discussed.

2.2 Communication Technologies

1. IEEE 802 Series

IEEE 802 is a family of standards that were developed to support Local Area Networks (LANs). For the Smart Grid illustrated in Fig. 6-2 (omitted), IEEE 802 standards are applicable to LANs in SCADA systems, NANs around the distribution networks and HANs in consumers' premises.

Fig. 6-3 (omitted) shows how the IEEE 802 architecture relates to the lowest two layers of the ISO/OSI reference model. It shows how two LANs may be connected through a Bridge. Such a connection is common in many organisations which have multiple LAN. A packet from the Source enters the Logical Link control (LLC) sublayer which acts as an interface between the network layer and the MAC sublayer. The LLC sublayer is defined by IEEE 802.2 and provides multiplexing mechanisms, flow control and error control. The packet then passes into the MACsublayer. At the MAC sublayer, a header and a trailer (depending on the LAN which the packet is entering) are added to the packet. Then it goes through the physical layer and the communication channel and reaches the Bridge. At the MAC layer of the Bridge, the header and trailer are removed and the original packet is recovered and passes to the LLC sublayer of the Bridge. Then the packet is processed (by adding an appropriate header and trailer) for the LAN to which it is forwarded (to the Destination) by the MAC layer. This use of a Bridge is essential as different LANs use different frame lengths and speeds. For example, IEEE 802.3 uses a frame of 1500 bytes whereas IEEE 802.4 uses a frame of 8191 bytes.

2. Mobile Communications

Mobile communication systems were designed initially to carry voice only. The standard that has enabled this technology is GSM (Global System for Mobile Communications). As an add-on data service to GSM technology, the General Packet Radio Service (GPRS) was developed. GPRS uses the existing GSM network and adds two new packet-switching network elements: the Gateway GPRS Support Node (GGSN) and the Serving GPRS Support Node (SGSN).

In December 1998, the European Telecommunications Standards Institute, the Association of Radio Industries and Businesses/Telecommunication Technology Committee of Japan, the China Communications Standards Association, the Alliance for Telecommunications Industry Solutions (North America) and the Telecommunications Technology Association (South Korea) launched a project called the 3rd Generation Partnership Project (3GPP). The aim of the 3GPP project was to develop a 3rd generation mobile systems (3G) based on GSM, GPRS and Enhanced Data Rates for GSM Evolution (EDGE). The project was built on data communication rather than voice. This project rapidly evolved to provide many different technologies as shown in Fig. 6-4 (omitted). The data rates of the different technologies that evolved under 3GPP are shown in Tab. 6-1 (omitted).

LTE is a competing technology to WiMax and supports user mobility up to 350 km/h, coverage up to 100 km, channel bandwidth up to 100 MHz with spectral efficiency of the Downlink 30 bps/Hz and the Uplink 15 bps/Hz. LTE has the advantage that it can support seamless connection to existing networks, such as GSM and UMTS as shown in Fig. 6-5 (omitted).

3. Multi Protocol Label Switching

Multi Protocol Label Switching (MPLS) is a packet forwarding technique capable of providing a Virtual Private Network (VPN) service to users over public networks or the internet. VPN provides the high quality of service and security required by Applications such as that associated with critical assets. Some anticipated Applications of point-to-point VPNs based on MPLS include Remote Terminal Unit (RTU) networks and backbone network to the System Control Centre. MPLS-based VPN is an attractive solution for wide area connectivity due to its relatively low cost and ability to be implemented rapidly using the existing networks resources. MPLS works by attaching labels to data packets received from the Network layer as shown in Fig. 6-6 (omitted). A MPLS header consists of four fields, namely: the 20-bit label field, the 3-bit experimental or class of service field, the stack bit and the 8-bit time to live field as shown in Fig. 6-7 (omitted). In MPLS, when a packet is forwarded, the label is sent with it to the next node. At that node, the label is used to determine the next hop. The old label is replaced with a new label and the packet is forwarded to its next hop.

4. Power Line Communication

(1) IEEE P1901

Under the sponsorship of the IEEE Communication Society, the IEEE P1901 working group was formed in 2005 with the remit to develop a standard for high speed (>100 Mbps at the physical layer) communication devices via electric power lines, the so-called Broadband over Power Line (BPL) devices. This project is devoted to producing a standard for BPL networks. The in-home and access protocol under IEEE P1901 will support MAC layer and Physical layers that use orthogonal frequency multiplexing (OFDM).

The standard which was in draft form at the end of 2010 will use transmission frequencies below 100 MHz and support BPL devices used for the first-mile/last-mile connections as well as BPL devices used in buildings for LANs and other data distribution. It ensures that the EMC limits set by national regulators are met so that it is compatible with other wireless and telecommunications systems.

(2) HomePlug

HomePlug is a non-standardized broadband technology specified by the HomePlug Powerline Alliance, whose members are major companies in communication equipment manufacturing and in the power industry.

HomePlug Powerline Alliance defines the following standards:

① HomePlug 1.0: connects devices in homes (1-10 Mbps).

② HomePlug AV and AV2: transmits HDTV and VoIP in the home − 200 Mbps (AV) and 600 Mbps (AV2).

③ HomePlug CC: Command and Control to complement other functions.

④ HomePlug BPL: still a working group addressing last-mile broadband (IEEE P1901).

The transmission technology OFDM used by HomePlug, is specially tailored for use in power line environments. It uses 84 equally spaced subcarriers in the frequency band between 4.5 and 21 MHz. Impulsive noise events common in power line environments are overcome by means of forward error correction and data interleaving.

Specialized English Words

communication technology　通信技术
System Control Centre　系统控制中心
operational facility　运营设施
Smart Grid　智能电网
substation　变电站
mobile communication　移动通信
power system　电力系统
Wide Area Network (WAN)　广域网
central generating station　中央发电站
communication infrastructure　通信设备
Home Area Network (HAN)　局域网
power line communication　电力线通信

transmission grid substation　输电网变电站
primary distribution substation　一次配电变电站
core communication network　核心通信网络
Global System for Mobile Communication（GSM）　全球移动通信系统
General Packet Radio Service（GPRS）　通信分组无线业务
Gateway GPRS Support Node（GGSN）　网关 GPRS 支持节点
Serving GPRS Support Node（SGSN）　GPRS 服务支持节点
Supervisory Control and Data Acquisition（SCADA）　监控与数据采集

Unit 3　Intelligent Sensor Technology

3.1 Definition of Smart Sensors

If we combine a sensor, an analog interface circuit, an analog to digital converter (ADC) and a bus interface in one housing, we get a smart sensor. Three hybrid smart sensors are shown in Fig. 6-8 (omitted), which differ in the degree to which they are already integrated on the sensor chip. This calls for standardization. And hence the sensor must become smarter.

In the first hybrid smart sensor, a universal sensor interface (USI) can be used to connect the sensor with the digital bus. In the second one, the sensor and signal conditioner have been integrated. However, the ADC and bus interface are still outside. In the third hybrid, the sensor is already combined with an interface circuit on one chip that provides a duty cycle or bit stream. Just the bus interface is still needed separately.

At this level, many output formats still exist, as shown in Tab. 6-2 (omitted).

3.2 Definition of Integrated Smart Sensors

If we integrate all functions from senior to bus interface on one chip, we get an integrated smart sensor, as depicted in Fig. 6-9 (omitted).

An integrated smart sensor should contain all elements necessary per node: one or more sensors, amplifiers, a chopper and multiplexers, an A/D converter, buffers, a bus interface, addresses, and control and power management. This is shown in Fig. 6-10 (omitted).

Although fully integrating all functions will be expensive, mass-production of the resulting sensor can keep the cost per integrated smart sensor reasonable. Another upside is that the costs of installing the total sensor system can be drastically reduced because of the simple modular architecture.

However, for realizing all functions on one chip we must first integrate a diversity of sensors on one chip. For this purpose an IC-compatible three-dimensional microstructuring technology is being developed. Tab. 6-3 (omitted) contains a number of IC-compatible sensors presently being developed.

In addition, interface electronics have to be developed, suitable for integration on the sensor chip. Tab. 6-4 (omitted) contains some examples of integrated smart sensors with on-chip interface electronics.

3.3 Definition of Integrated Smart Sensor Systems

Fig. 6-11 (omitted) depicts the evolution of integrated smart sensor systems with many intermediate steps. The greater the market for smart sensors of a certain type, the more integration is economically affordable for that type.

Our final dream is depicted in Fig. 6-12 (omitted). If we are also able to integrate a wireless power source and wireless communication, a whole new concept of ambiguous sensors will appear. Many sensors could then be used in cars, homes, clothes, and fields to obtain valuable information.

3.4 Object-Oriented Design of Sensor Systems

System performance can be improved significantly and the cost can be reduced by merging and reevaluating the functions of sensors, actuators, analog interfacing circuits and digital processors in overall designs. Where technology allows, the system can be implemented on a common substrate or in a single-chip integrated circuit.

In this unit we will consider sensor systems that are targeted at a cost-driven, medium-volume, industrial sensor market. According to Toth, when designing sensor systems, the traditional top-down and bottom-up design approaches have serious limitations. These limitations are due to the interdisciplinary and relatively open character of the sensor subsystems. Consequently, the traditional design methods often require too many iteration steps and result in a long design time and inflexible designs. To overcome these limitations, system designs and specifications should be reused as much as possible. By analogy with a similar design approach in software engineering, Toth refers to this approach as the Object-Oriented Design. Fig. 6-13 (omitted) shows a possible hardware configuration resulting from such an approach for a sensor system in which powerful components have been used, such as microcontrollers (uCs), personal computers (PCs) and sensor interfaces. The availability of memory in the microcontroller makes it possible to collect data over a longer period for a number of sensors. This enables the realization of several important system functions, such as autocalibration, self-testing and the compensation and filtering of undesired signals and effects. As will be explained, the A/D conversion can be performed in the microcontroller. When the sensor signal is converted to the time domain, using a period-modulator in the transducer interface, microcontrollers can perform this task very well, even without using a built-in A/D converter.

The transducer interface is equipped with front-end electronics for various types of sensors. Sometimes, but not always, it is possible to merge the functions of the sensing element and its interface and to implement them on a single chip, resulting in a so-called “Smart Sensor” (Fig. 6-14 (omitted)). In any case, the electrical properties of the front-end electronics should exactly match those of the sending elements, taking into account the specific properties, circumstances and nonidealities. In designing a match, object-oriented design will help to speed up design and save costs. To enable this, it is important to recognize the main features and problems of the most common types of sensing elements and sensor systems.

Specialized English Words

digital bus　数字总线
bit stream　比特流
multiplexer　多工器/多路器
actuator　执行器
wireless power source　无线电源
substrate　基板；衬底
autocalibration　自动校准
transducer interface　传感器接口
duty cycle　占空比
chopper　斩波器
buffer　缓冲器
analog interfacing circuit　模拟接口电路
digital processor　数字处理器
single-chip integrated circuit　单片集成电路
self-testing　自校验
smart sensor　智能传感器
universal sensor interface (USI)　通用传感器接口
intelligent sensor technology　智能传感技术
analog-to-digital converter　模/数转换器

Unit 4 Smart Grid Information Technology

4.1 Introduction

The operation of a Smart Grid relies heavily on two-way communication for the exchange of information. Real-time information must flow all the way to and from the large central generators, substations, customer loads and the distributed generators. At present, power system communication systems are usually restricted to central generation and transmission systems with some coverage of high voltage distribution networks. The generation and transmission operators use private communication networks, and the SCADA and ICT systems for the control of the power network are kept separate even from business and commercial applications operated by the same company. Such segregation of the power system communication and control system (using private networks and proprietary control systems) limits access to this critical ICT infrastructure and naturally provides some built-in security against external threats.

With millions of customers becoming part of the Smart Grid, the information and communication infrastructure will use different communication technologies and network architectures that may become vulnerable to theft of data or malicious cyber attacks. Ensuring information security in the Smart Grid is a much more complex task than in conventional power systems because the systems are so extensive and integrated with other networks. Potentially sensitive personal data is transmitted and, in order to control costs, public ICT infrastructure such as the Internet will be used.

Obtaining information about customers' loads could be of interest to unauthorised persons and could infringe the privacy of customers. The ability to gain access to electricity use data and account numbers of customers opens up numerous avenues for fraud. Breaching the security of power system operating information by an unauthorised party has obvious dangers for system operation.

The Smart Grid requires reliable and secure delivery of information in real time. It not only needs throughput, the main criterion adopted to describe performance required for common internet traffic. Delays in the accurate and safe delivery of information are less tolerable in the Smart Grid than for much commercial data transmission as the information is required for real-time or near real-time monitoring and control. A lot of monitoring and control information is periodic and contributes to a regular traffic pattern in the communication network of the Smart Grid, though during power system faults and contingencies there will be a very large number of messages. The length of each message will be short. The pattern of messages passed in the ICT system of the Smart Grid is very different from that of a traditional voice telephone system or the Internet. Any form of interruption resulting from security issues is likely to have serious effects on the reliable and safe operation of the Smart Grid.

Security measures should ensure the following:

(1) Privacy that only the sender and intended receiver(s) can understand the content of a message.

(2) Integrity that the message arrives in time at the receiver in exactly the same way it was sent.

(3) Message authentication that the receiver can be sure of the sender's identity and that the message does not come from an imposter.

(4) Non-repudiation that a receiver is able to prove that a message came from a specific sender and the sender is unable to deny sending the message.

Providing information security has been a common need of ICT systems since the Internet became the main mode of communication. Thus there are well-established mechanisms to provide information security against possible threats.

4.2 Encryption and Decryption

Cryptography has been the most widely used technique to protect information from adversaries. As shown in Fig. 6-15 (omitted), a message to be protected is transformed using a Key that is only known to the Sender and Receiver. The process of transformation is called encryption and the message to be encrypted is called Plain text. The transformed or encrypted message is called Cipher text. At the Receiver, the encrypted message is decrypted.

Fig. 6-15 (omitted) also shows possible threats to a message. As indicated in Fig. 6-15 (omitted), an intruder can launch a passive or an active attack. A passive attack may use captured information for malicious purposes. In an active attack, data may be modified on its path or completely new data may be sent to the Receiver. Passive attacks, though they do not pose an immediate threat, are hard to detect. Active attacks are more destructive but can be detected quickly in most situations.

1. Symmetric Key Encryption

In classical encryption both the sender and receiver share the same Key. This is called symmetric key encryption.

(1) Substitution Cipher

Substitution cipher was an early approach based on symmetric Key encryption. In this process, each character is replaced by another character. An example of a mapping in a substitution cipher system is shown in Tab. 6-1 below:

Tab. 6-1 The mapping table of original and decoding code

Plain text	A	B	C	D	E	F	G	H	I	J	K	L	M	N	O	P	Q	R	S	T	U	V	W	X	Y	Z
Cipher text	W	Y	A	C	Q	G	I	K	M	O	E	S	U	X	Z	B	D	F	H	J	L	N	P	R	T	V

The encryption of message or plain text HELLO THERE will produce KQSSZ JKQFQ as Cipher text. Since a given character is replaced by another fixed character, this system is called a mono-alphabetic substitution. The Key here is the string of 26 characters corresponding to the full alphabet. Substitution cipher systems disguise the characters in the Plain text but preserve the order of characters in the Plain text.

Even though the possible Key combinations, $26! \approx 4\times10^{26}$, appear to be large enough that this system feels safe, by using the statistical properties of natural languages, the Cipher text can easily be broken.

The possibility of exploiting the statistical properties of natural languages, due to mono-alphabetic substitution, can be reduced by the use of polyalphabetic substitution, in which each occurrence of a character can have a different substitute. In other words, the relationship between a character in Plain text and a character in Cipher text is one-to-many. For instance, the substituting character corresponding to a character can be made to depend upon the relative position of the character within the Plain text.

(2) Transposition Cipher

In a transposition cipher the characters in the Plain text are transposed to create the Cipher text. Transposition can be achieved by organising the Plain text into a two-dimensional array and interchanging columns according to a rule defined by a Key. An example of transposition cipher is shown in Fig. 6-16 (omitted). As can be seen, Plain text is first assigned to an array having the same number of columns as the Key. Any unused columns are filled with the letter “a”. Then each row in the array is rearranged in the alphabetical order of the Key.

(3) One-Time Pad

In the one-time pad method a random bit string of the same length as the message is combined using the bit-wise exclusive OR (XOR) with the Plain text. The adversary has no information at all for breaking the Cipher text produced by a one-time pad since every possible Plain text is equally probable.

In this method, both the Sender and Receiver have to carry the same random bit string

Key of encryption. This is not an easy task as a Key can only be used once and the amount of data that can be transferred with a given Key is limited by the length of the Key.

If a one-time pad Key is used to encrypt more than one message, this can cause vulnerabilities in security. To understand this better, consider a situation where plain text message M_1 and M_2 are encrypted with the same Key K. Let C_1 and C_2 be the corresponding Cipher texts:

$$C_1 = K \oplus M_1$$
$$C_2 = K \oplus M_2$$

Hence,

$$C_1 \oplus \bar{C}_2 = M_1 \oplus M_2$$

Therefore, if an adversary gets to know either M_1 or M_2 the other message can be computed easily.

(4) Data Encryption Standard

The Data Encryption Standard (DES) is a typical example of modem cryptography. As a well-engineered encryption standard, it or its variants have been in widespread use since the early 1970s. In DES, blocks of data having 64 bits are treated as units for encryption. Fig. 6-17 (omitted) shows the DES algorithm. As shown in Fig. 6-17 (a) (omitted), the Plain text is encrypted 16 times using 48-bit subkeys (K_1, ..., K_{10}) generated by a 56-bit key. The F-function indicated in Fig. 6-17 (a) (omitted) is the Feistel function and shown in Fig. 6-17 (b) (omitted).

Initial permutation and final permutation are used to rearrange the input and output bits using a mapping method. The duplication and expansion box is used to expand the right 32 bits of the plain text to 48-bits after initial permutation. The expanded bits are then combined with the 48-bit subkey, K_j, using the XOR operation. The output is then grouped into 6-bit groupings. These groupings are sent through S-boxes which use an address to locate an output using another mapping matrix. Each S-box uses different mapping matrices and provides 4-bit output. Each output is then combined to form a 32-bit stream and that goes through a P-box (using a mapping matrix as the initial permutation to rearrange the bits) to form the output of each F-function.

There have been many approaches developed to break the DES cipher, yet the best practical attack known is the exhaustive Key search. Concerns over the relatively short Key length of 56 bits led to the DES being strengthened in a simple way that led to the development of Triple DES. The Triple DES ciphers use three iterations, encryption-decryption-encryption, of DES as shown in Fig. 6-18 (omitted). The DES decryption is achieved by using the same 16-round process but with subkeys used in reverse order, K_{10} through K_1.

Apart from DES, there are several other block cipher algorithms in common use. Examples of commonly used algorithms are: Advanced Encryption Standard (AES), Blowfish, Serpent and Twofish.

Some applications find that a block cipher has drawbacks. For instance, since block ciphers need to have a complete block of data to commence processing, it makes both the encryption and decryption processes slow. A stream cipher is another symmetric key cipher where Plain text bits are typically XOR with a pseudorandom Key stream. Stream cipher can be considered a one-time pad encryption. RC4 is an example of a commonly used stream cipher algorithm.

2. Public Key Encryption

Key distribution is an issue for all cryptography. Symmetric key encryption algorithms require a secure initial exchange of secret Keys between the sender and receiver and the number of secret Keys required grows with the number of devices in a network. Public Key encryption does not require secure initial exchanges of secret Keys between the sender and receiver.

Public key algorithms involve a pair of keys called the public Key and the private Key. Each user announces its public Key but retains its private Key confidentially. If user A wishes to send a message to user B, then A encrypts the message using B's public Key. Public Key algorithms are such that it is practically not possible to determine the decryption Key even though the encryption Key is known as it uses one key for encryption and another for decryption. RSA (this acronym stands for Rivest, Shamir and Adleman who first publicly described it) is a widely used public Key algorithm.

The RSA algorithm requires a methodical generation of Keys. The following is the process:

① Choose two distinct prime numbers p and q.

② Compute $n = p \times q$.

③ Compute $z = (p-1)(q-1)$.

④ Choose an integer e such that it is relatively prime to z (both e and z have no common factors).

⑤ Find d such that $e \times d = 1 \pmod{z}$ (in other words $e \times d - 1$ should be an integer multiple of z).

⑥ The pair (e, n) is then released as the public Key and pair (d, n) is used as the private Key.

⑦ The encryption of message M gives the Cipher text $C = M' \times \mathrm{Mod}\ (n)$.

⑧ The decryption is done by computing $M = C' \times \mathrm{Mod}\ (n)$.

Example:

Let's consider that user B wishes to transmit character "D" to user A. The following is the steps involved in the RSA algorithm:

① A selects p and q. Assume that he chooses $p = 11$ and $q = 3$, p and q are prime numbers.

② A computes $n = p \times q = 33$.

③ A computes $z = (p - 1) \times (q - 1) = 20$.

④ A chooses $e = 7$ which is relatively prime to $z = 20$.

⑤ A finds $d = 3$ that makes $3 \times 7 = 21 \equiv 1 \pmod{20}$

⑥ A publishes pair (7, 33) as the public Key.

⑦ B uses A's public Key (33, 7) and computes $C = 4^7$ Mod 33 = 16 and transmits 16 to A (D is the 4th character in the alphabet).

⑧ A computes M = 163 Mod 33 = 4 (A obtains D).

4.3 Authentication

Authentication is required to verify the identities of communicating parties to avoid imposters gaining access to information. When user A receives a communication from user B, A needs to verify that it is actually B, but not someone else masquerading as B, who is talking to him.

(1) Authentication Based on Shared Secret Key

Assume that A and B wish to establish a communication session. Prior to exchanging data, they need to authenticate each other. The steps involved in this method are shown in Fig. 6-19 (omitted).

(2) Authentication Based on Key Distribution Centre

This method involves a trusted key distribution centre (KDC) that supports the authentication. KDC and each of its users have a shared secret Key that is used to communicate between them. The steps involved are shown in Fig. 6-20 (omitted).

Specialized English Words

information technology　信息技术
generator　发电机
substation　变电站
encryption　加密
decryption　解密
cryptography　密码学
intruder　入侵者
symmetric key encryption　对称金钥加密
substitution cipher　替代密码
transposition cipher　转置密码，移位密码
one-time pad　单次秘本
Data Encryption Standard (DES)　数据加密标准
cipher　密码
public key encryption　公钥加密
key distribution centre (KDC)　密钥分配中心

Unit 5 Intelligent Power Distribution Technology and Application

5.1 Introduction

Modern electric power systems are supplied by large central generators that feed power into a high voltage interconnected transmission network. The power, often transmitted over long distances, is then passed down through a series of distribution transformers to final circuits for delivery to customers (Fig. 6-21 (omitted), also refer to Plate 1).

Operation of the generation and transmission systems is monitored and controlled by Supervisory Control and Data Acquisition (SCADA) systems. These link the various elements through communication networks (for example, microwave and fibre optic circuits) and connect the transmission substations and generators to a manned control centre that maintains system security and facilitates integrated operation. In larger power systems, regional control centres serve an area, with communication links to adjacent area control centres. In addition to this central control, all the generators use automatic local governor and excitation control. Local controllers are also used in some transmission circuits for voltage control and power flow control, for example, using phase shifters (sometimes known as quadrature boosters).

Traditionally, the distribution network has been passive with limited communication between elements. Some local automation functions are used such as on-load tap changers and shunt capacitors for voltage control and circuit breakers or auto-reclosers for fault management. These controllers operate with only local measurements and wide-area coordinated control is not used.

Over the past decade, automation of the distribution system has increased in order to improve the quality of supply and allow the connection of more distributed generation. The connection and management of distributed generation are accelerating the shift from passive to active management of the distribution network. Network voltage changes and fault levels are increasing due to the connection of distributed generation. Without active management of the network, the costs of connection of distributed generation will rise and the connection of additional distributed generation may be limited.

The connection of large intermittent energy sources and plug-in electric vehicles will lead to an increase in the use of Demand-Side Integration and distribution system automation.

5.2 Substation Automation Equipment

The components of a typical legacy substation automation system are shown in Fig. 6-22 (omitted). Traditionally, the secondary circuits of the circuit breakers, isolators, current and voltage transformers and power transformers were hard-wired to relays. Relays were connected with multi-drop serial links to the station computer for monitoring and to allow remote interrogation.

However, the real-time operation of the protection and voltage control systems was through hard-wired connections.

The configuration of a modern substation automation system is illustrated in Fig. 6-23 (omitted). Two possible connections (marked by boxes) of the substation equipment are shown in Fig. 6-24(omitted). Although it may vary from design to design, generally it comprises three levels:

(1) The *station level* includes the substation computer, the substation human machine interface (which displays the station layout and the status of station equipment) and the gateway to the control centre.

(2) The *bay level* includes all the controllers and intelligent electronic devices (which provide protection of various network components and a real-time assessment of the distribution network).

(3) The *process level* consists of switchgear control and monitoring, current transformers (CTs), voltage transformers (VTs) and other sensors.

In connection 1, analog signals are received from CTs and VTs (1 A or 5 A and 110V) as well us status information and are digitized at the bay controller and IEDs. In connection 2, analog and digital signals received from CTs and VTs are digitized by the interfacing unit. The process bus and station bus take these digital signals to multiple receiving units, such as IEDs, displays, and the station computer that are connected to the Ethernet network. To increase reliability, normally two parallel process buses are used (only one process bus is shown in Fig. 6-23 (omitted)).

The station bus operates in a peer-to-peer mode. This is a LAN formed by connecting various Ethernet switches through a fibre-optic circuit. The data collected from the IEDs is processed for control and maintenance by SCADA software that resides in the station computer.

The hard-wiring of traditional substations required several kilometers of secondly wiring in ducts and on cable trays. This not only increased the cost but also made the design inflexible. In modern substations as inter-device communications are through Ethernet and use the same communication protocol, IEC 61350, both the cost and physical footprint of the substations have been reduced.

Specialized English Words

circuit breaker 断路器
isolator 隔离器
power transformer 电力变压器
relay 继电器
switchgear 开关设备
communication network 通信网络
current and voltage transformer 电流、电压互感器
intelligent power distribution technology 智能配电技术
high voltage interconnected transmission network 高压互联输电网络
supervisory control and data acquisition (SCADA) 监控与数据采集
substation automation equipment 变电站自动化设备
legacy substation automation system 传统变电站自动化系统

Unit 6　科技英语写作方法

英语科技论文写作是进行国际学术交流必需的技能。一般而言，发表在专业英语期刊上的科技论文在文章结构和文字表达上都有其特定的格式和规定。撰写英语科技论文的第一步就是推敲结构。

英语科技论文的基本格式包括：

Title——论文题目

Author(s)——作者姓名

Affiliation(s) and address(es)——联系方式

Abstract——摘要

Keywords——关键词

Body——正文

Acknowledgements——致谢（可空缺）

References——参考文献

Appendix——附录（可空缺）

Resume——作者简介（视刊物而定）

其中正文为论文的主体部分，分为若干章节。一篇完整的科技论文的正文部分由以下内容构成：

Introduction——引言/概述

Methods——方法

Results——结果

Discussion——讨论

Conclusions——结论/总结

下面对科技论文主要构成部分的写法和注意事项进行详细介绍。

6.1 Abstract（摘要）

摘要也称为内容提要，是对论文的内容不加注释和评论的简短陈述。其作用主要是为读者阅读、信息检索提供方便。摘要不宜太详尽，也不宜太简短，应将论文的研究体系、主要方法、重要发现、主要结论等简明扼要地加以概括。

摘要的构成要素：

（1）研究目的——准确描述该研究的目的，说明提出问题的缘由，阐明研究的范围和重要性。

（2）研究方法——简要说明研究课题的基本设计，阐明结论是如何得到的。

（3）结果——简要列出该研究的主要结果和新发现，说明其价值和局限。叙述要具体、准确并给出结果的置信值。

（4）结论——简要地说明经验，论证取得的正确观点及理论价值或应用价值，说明是否还有与此有关的其他问题有待进一步研究、是否可推广应用等。

6.2 Introduction（引言）

引言位于正文的起始部分，主要叙述写作的目的或研究的宗旨，使读者了解和评估研究成果。主要内容包括：介绍相关研究的历史、现状、进展，说明自己对已有成果的看法和以往工作的不足之处，以及自己所做研究的创新性或重要价值；说明研究中要解决的问题、所采取的方法，必要时须说明采用某种方法的理由；介绍论文的主要结果和结构安排。

6.3 Methods（方法）

在论文中，这一部分用于说明实验的对象、条件、使用的材料、实验步骤或计算过程、公式的推导、模型的建立等。对过程的描述要完整具体，符合其逻辑步骤，以便读者重复实验。

对方法的描述要详略得当、重点突出。应遵循的原则是给出足够的细节信息以便让同行能够重复实验，避免混入与结果或发现有关的内容。如果方法新颖，且不曾发表过，应提供所有必需的细节；如果所采用的方法已经公开报道过，引用相关的文献即可（如果报道该方法的期刊影响力很有限，可稍加描述）。

力求语法正确、描述准确。由于方法部分一般需要描述很多内容，因此通常需要采用很简洁的语言，故使用精确的英语描述材料和方法是十分重要的。在时态与语态的运用上，若描述的内容为不受时间影响的事实，采用一般现在时；若描述的内容为过去某一特定的行为或事件，则采用过去时；方法的焦点在于描述实验中所进行的每个步骤以及所采用的材料，由于所涉及的行为是讨论的焦点，而且读者已知道进行这些行为的人就是作者自己，因而一般采用被动语态。如果涉及表达作者的观点或看法，则应采用主动语态。

6.4 Results（结果）

本部分描述研究结果，它可自成体系，以便令读者不必参考论文其他部分，也能了解作者的研究成果。对结果的叙述也要按照其逻辑顺序进行，使之既符合实验过程的逻辑顺序，又符合实验结果的推导过程。本部分还可以包括对实验结果的分类整理和对比分析等。

对实验或观察结果的表达要高度概括和提炼，不能简单地将实验记录数据或观察事实堆积到论文中，尤其是要突出有科学意义和具代表性的数据，而不是没完没了地重复一般性数据。对实验结果的叙述要客观真实，即使得到的结果与实验不符，也不可略而不述，而且还应在讨论中加以说明和解释。数据表达可采用文字与图表相结合的形式。在时态的运用上，常用一般现在时。

6.5 Discussion（讨论）

"讨论"的重点在于对研究结果的解释和推断，并说明作者的结果是否支持或反对某种观点、是否提出了新的问题或观点等。因此撰写讨论时要避免含蓄，尽量做到直接、明确，以便审稿人和读者了解论文为什么值得引起重视。讨论的内容主要有：(1)回顾研究的主要目的或假设，并探讨所得到的结果是否符合原来的期望。如果没有的话，为什么？(2)概述最重要的结果，并指出其是否能支持先前的假设以及是否与其他学者的结果相一致。如果不是的话，为什么？(3)对结果提出说明、解释或猜测。根据这些结果，能得出何种结论或推论？(4)指出研究的限制以及这些限制对研究结果的影响，并提出对进一步的研究题目或方向的建议。(5)指出结果的理论意义(支持或反驳相关领域中的现有理论、修正现有理论)和实际应用。

6.6 Conclusions（结论）

作者在文章的最后要单独用一个章节对全文进行总结，其主要内容是对研究的主要发现和成果进行概括总结，让读者对全文的重点有一个深刻的印象。有的文章也在本部分提出当前研究的不足之处，对研究的前景和后续工作进行展望。应注意的是，撰写结论时不应涉及前文不曾指出的新事实，也不能在结论中重复论文其他章节中的句子，或者叙述其他不重要或与自己研究没有密切联系的内容，以故意把结论拉长。

实训课程六：论文摘要的撰写

摘要是一篇论文的缩影，更是一篇文章的精华。因此，摘要既要高度概括论文，使其易于阅读，更要阐明论文的重点和亮点。通常，一篇摘要包括以下内容：

(1)目的：研究工作的目的或研究工作希望解决的问题。在言明研究目的之前，往往会以研究背景作为引言。在时态的运用上：如果研究背景是不受时间影响的普遍事实，应使用现在时；如果研究背景是对某种研究趋势的概述，则使用现在完成时。如果研究目的是“以论文为导向”，多用现在时，如“This paper focuses on ...”；如果研究目的是“以研究为导向”，则多用过去时，如“This work investigated ...”。

（2）方法：所用的理论、模型、算法等，即如何解决问题。在时态的运用上，多使用现在时，如：“A ... approach is applied to ...”。

（3）结果：实验的结果。在时态的运用上，通常用现在时。

（4）结论：对研究的问题或实验结果的分析、比较和评价。在时态的运用上，通常用一般现在时。

下面以下文为例，介绍科技论文摘要的撰写方式。

Abstract: Most of the existing demand-side management programs focus primarily on the interactions between a utility company and its customers/users. In this paper, we present an autonomous and distributed demand-side energy management system among users that takes advantage of a two-way digital communication infrastructure which is envisioned in the future smart grid. We use game theory and formulate an energy consumption scheduling game, where the players are the users and their strategies are the daily schedules of their household appliances and loads. It is assumed that the utility company can adopt adequate pricing tariffs that differentiate the energy usage in time and level. We show that for a common scenario, with a single utility company serving multiple customers, the global optimal performance in terms of minimizing the energy costs is achieved at the Nash equilibrium of the formulated energy consumption scheduling game. The proposed distributed demand-side energy management strategy requires each user to simply apply its best response strategy to the current total load and tariffs in the power distribution system. The users can maintain privacy and do not need to reveal the details on their energy consumption schedules to other users. We also show that users will have the incentives to participate in the energy consumption scheduling game and subscribing to such services. Simulation results confirm that the proposed approach can reduce the peak-to-average ratio of the total energy demand, the total energy costs, as well as each user's individual daily electricity charges.（摘自论文“Autonomous Demand-Side Management Based on Game-Theoretic Energy Consumption Scheduling for the Future Smart Grid”）

该摘要的引言部分，即“Most of the existing demand-side management programs focus primarily on the interactions between a utility company and its customers/users.”，采用了“Most of the existing ... focus on ...”的句型回顾论文的研究背景。常用句型还有“Much/Little work/research has been carried out on ...”“Many/Quite few studies has been done regarding ...”等。

“In this paper, we present an autonomous and distributed demand-side energy management system among users that takes advantage of a two-way digital communication infrastructure which is envisioned in the future smart grid.”阐明了该论文的研究目的。常

用词汇有 purpose、attempt、aim、objective 等。另外还可以用动词不定式充当目的状语来表达，如 in an attempt to、in order to、to investigate 等。

“We use game theory ... We also show that users. .. such services.”介绍了论文采用的研究方法。常用的词汇有 use、apply、employ、utilize、adopt、investigate、discuss、analyze、show、present 等。

最后，“Simulation results confirm that the proposed approach ..., as well as each user's individual daily electricity charges.”展示了论文的研究结果。常用词汇有use、apply、application等。常用的句型有“The results、studies、test showed、demonstrated、illustrated、suggested that ...”等。

此外，英语摘要一般不用数学公式。有一种说法是，英语摘要通常使用第三人称、被动语态，这样可以弱化作者，强化信息，但实际上在有些权威期刊中，有些论文摘要也出现了第一人称、主动语态。可见，是否采用第三人称、被动语态可以根据实际情况而定，不一定要刻意回避使用第一人称、主动语态。

附录：专业英语缩写词汇表

A

ABB Air Arcing Blaster 空气吹弧断路器
ABM Asynchronous Balanced Mode 异步平衡方式
AC Alternating Current 交流电
ACB Air Circuit Breaker 空气断路器
ACS Alternating Current Synchronous 交流同步
ADAC Analog-Digital-Analog Converter 模拟数字模拟转换器
ADC Analog-to-Digital Converter 模拟/数字转换器
AFC Automatic Following Control 自动跟踪控制
AM Amplitude Modulation 调幅
AMP Amplifier 放大器
AO Analog Output 模拟输出
AOM Analog Output Module 模拟输出模块
APC Automatic Phase Control 自动相位控制
APD Avalanche Photo Diode 雪崩光电二极管
APFR Automatic Power Factor Regulator 自动功率因数调节器
APFC Active Power Factor Correction 有源功率因数校正
APS Automatic Phase Shifter 自动移相器
APS Automatic Phase Synchronization 自动相位同步
AR Automatic Reclose 自动重合闸
ARC Auto Reclose 自动重合闸
ARC Automatic Remote Control 自动遥控
ARM Asynchronous Response Mode 异步响应方式
ASC Automatic Signal Converter 模拟信号转换器
AVQC Automatic Voltage and Reactive Power Control 自动电压无功控制

AVR Automatic Voltage Regulator 自动电压调整器
AZS Automatic Zero Setting 自动调零

B

BIL Basic Impulse Insulation Level 基本冲击绝缘水平
BIL Basic Impulse Level 基本脉冲水平
BIFET Bipolar Field-Effect Transistor 双极场效应晶体管
BIM Bus Interface Module 总线接口模块
BIOS Basic Input Output System 基本输入 / 输出系统
BL Bay Level 间隔层
BV Breakdown Voltage 击穿电压
BJT Bipolar Junction Transistor 双极结型晶体管

C

CB Circuit Breaker 断路器
CB Control Board 控制盘
CB Control Bus 控制总线
CBR Constant Bit Rate 固定比特率
CC Constant Current 恒流
CCM Continuous Conduction Mode 电流连续模式
CCD Charge-Coupled Device 电荷耦合器件
CD Carrier Detect 载波检测
CL Current Loop 电流环路
CMOS Complementary Metal Oxide Semiconductor 互补金属氧化物半导体
CO Control Output 控制输出
CR Controlled Rectifier 可控整流器
CPU Central Processing Unit 中央处理单元
CS Control Signal 控制信号
CS Control Switch 控制开关
CSF Control System Function Chart 控制系统功能图解
CTs Current Transformers 电流互感器
CSTI Current Source Type Inverter 电流（源）型逆变电路
CVCF Constant Voltage Constant Frequency 恒压恒频

D

DAC Digital to Analog Converter 数/模转换器
DAR Dielectric Absorption Ratio 介质吸收比
DAS Distribution Automation System 配电自动化系统
DC Direct Current 直流电
DCS Distributed Control System 分布式控制系统
DCS Digital Control System 数字控制系统
DCM Discontinuous Conduction Mode 电流断续模式
DDC Direct Digital Control 直接数字控制
DF Distortion Factor 失真系数
DI Digital Input 数字输入
DIAC Diode Alternating Current Switch 二极管交流开关
DID Digital Information Display 数字信息显示
DO Digital Output 数字输出
DOB Data Output Bus 数据输出总线
DP Data Processing 数据处理
DPS Data Processing System 数据处理系统
DPU Decentralized Process Unit 分散处理单元
DSC Digital Signal Converter 数字信号转换器
DSP Digital Signal Processor 数字信号处理器
DSP Digital Signal Processing 数字信号处理
DSS Digital Satellite System 数字卫星系统
DMC Dynamic Motion Control 动态控制
DTU Distribution Terminal Unit 配电终端单元
DVR Dynamic Voltage Restorer 动态电压恢复器

E

EH Equivalent Hours 等效（运行）小时
EHV Extra High Voltage 超高压
EMC Electromagnetic Compatibility 电磁兼容性
EMF Electromotive Force 电动势
EMF Electromagnetic Force 电磁力
EMI Electromagnetic Interference 电磁干扰
EMS Energy Management System 能量管理系统

EMS Energy Metering System 电能计量系统
EPS Emergency Power Supply 应急电源
ERV Electrical Release Valve 电气释放阀
ESD Electrostatic Discharge 静电放电
ESDA Electronics System Design Automation 电子系统设计自动化

F

FACTS Flexible AC Transmission System 柔性交流输电系统
FC Fuse Contactor 熔断式接触器
FCB Fast Cut Back 快速切负荷
FCS Field Bus Control System 现场总线控制系统
FCT Field Controlled Thyristor 场控晶闸管
FEMF Foreign Electro-Motive Force 外部电动式
FET Field Effect Transistor 场效应晶体管
FEP Front-End Processor 前端处理器
FFC Flat Frequency Control 恒定频率控制
FM Frequency Modulation 调频
FST Fast Switching Thyristor 快速晶闸管
FRA Frequency Response Analysis 频响分析
FRD Fast Recovery Diode 快恢复二极管
FRED Fast Recovery Epitaxial Diode 快恢复外延二极管
FTC Flat-Tie line Control 联络线定潮流控制
FTU Feeder Terminal Unit 馈线终端单元
FVC Frequency-to-Voltage Converter 频率电压转换器

G

GCS Gate Controller Switch 可关断晶闸管；门控开关
GIS Gas Insulated Switchgear 气体绝缘开关设备
GTO Gate Turn-Off 闸门电路断开
GTR Giant Transistor 电力晶体管
GPS Global Positioning System 全球定位系统

H

HFC High Frequency Current 高频电流
HFO High Frequency Oscillator 高频振荡器
HRI_n Harmonic Ratio for I_n 次谐波电流含有率
HI-POT High Potential 高电位
HTSC High-Temperature Superconductor 高温超导体
HV High Voltage 高电压
HVDC High Voltage Direct Current 直流高压
HVIC High Voltage Integrated Circuit 高压集成电路

I

IC Integrated Circuit 集成电路
ICC Integrated Component Circuit 集成元件电路
ICE Integrated Circuit Emulator 集成电路仿真器
ICP Integrated Circuit Package 集成电路组件
IFT Intermediate Frequency Transformer 中频变压器
IEE Institute of Electrical Engineers 电气工程师学会
IEEE Institute of Electrical and Electronic Engineers 电气及电子工程师学会
IPB Isolated-Phase Bus 分相封闭式母线
IPC Interphase Power Controller 相间功率控制器
IPM Intelligent Power Module 智能功率模块
IR Infrared Rays 红外线
IR Insulation Resistance 绝缘电阻
IRED Infrared Emitting Diode 红外发光二极管
IGBT Insulated Gate Bipolar Transistor 绝缘栅双极晶体管
IGFET Insulated Gate Field Effect Transistor 绝缘栅场效应晶体管
IGCT Integrated Gate-Commutated Thyristor 集成门极换流晶闸管

J

JEDEC Joint Electronic Device Engineering Council 电子器件工程联合会
JFET Junction Field Effect Transistor 连结式场效应管

L

LA Lighting Arrester 避雷器
LASER Light Activated Silicon Controlled Rectifier 光激发可控硅整流器
LCD Liquid Crystal Display 液晶显示
LDR Light-Dependent Resistors 光敏电阻
LED Light Emitting Diode 发光二极管
LFC Load Frequency Control 负荷频率控制
LIC Linear Integrated Circuit 线性集成电路
LLC Logical Link Control 逻辑链路控制
LN Logical Node 逻辑节点
LRT Load Ratio Transformer 带负载切换分接头变压器
LS Level Switch 电平开关
LS Limit Switch 极限开关
LSB Least Significant Bit 最低有效数位
LSI Large-Scale Integrated Circuit 大规模集成电路
LSD Least Significant Digit 最小有效数字
LT Line Termination 线路终端
LTT Light Triggered Thyristor 光控晶闸管
LV Low Voltage 低电压

M

MCB Micro Circuit Breaker 微型断路器
MCR Maximum Capacity Rating 最大额定容量
MCS Modulating Control System 模拟量控制系统
MCT MOS Controlled Thyristor MOS 控制晶闸管
MMF Magnetomotive force 磁动势
MOA Metal-Oxide Arrester 金属氧化物避雷器
MOP Motorized Potentiometer 电动式电位计
MOS Metal-Oxide Semiconductor 金属氧化物半导体
MOSIC Metal-Oxide Semiconductor Integrated Circuit 金属氧化物半导体集成电路
MOST Metal-Oxide Semiconductor Transistor 金属氧化物半导体晶体管
MOST Metal-Oxide Silicon Transistor 金属氧化硅晶体管
MOV Metal-Oxide Varistor 金属氧化物可变电阻器
MTSV Master Trip Solenoid Valve 主跳闸电磁阀

MUF Maximum Useable Frequency 最大可用频率
MV Medium Voltage 中压

N

NC Numeric Control 数字控制
NCC Normally Closed Contact 常闭接点

O

OC Open Collector 集电极开路
OCB Oil Circuit Breaker 油断路器
OCT Optical Current Transducer 光学电流传感器
OFT Optical Fiber Tube 光导纤维管
OLC Open Loop Control 开环控制
OLTC On-Load Tap Changer 有载分接开关
ONAF Oil-Immersed Natural Air Forced 油浸强迫风冷
ONAN Oil-Immersed Natural Air Natural 油浸自然风冷
Op-Amp Operational Amplifier 运算放大器
OPC Optical Photoconductor 光导电体
OVT Optical Voltage Transducer 光学电压传感器

P

PA Power Amplifier 功率放大器
PAM Pulse Amplitude Modulation 脉冲振幅调制
PCB Printed Circuit Board 印刷电路板
PCM Pulse Code Modulation 脉码调制
PIO Process Input Output 过程量输入输出
PF Power Factor 功率因数
PFC Power Factor Correction 功率因数校正
PFM Pulse Frequency Modulation 脉冲频率调制
PIC Power Integrated Circuit 功率集成电路
PLC Power Line Carrier 输电线载波
PLC Programmable Logic Cell 可编程逻辑单元
PLC Programmable Logic Controller 可编程逻辑控制器

PLD Programmable Logic Device 可编程逻辑器件
PLIC Programmable Logic Integrated Circuit 可编程逻辑集成电路
PMOS Positive-Channel Metal Oxide Semiconductor P 通道金属氧化物半导体
PM Phase Modulation 调相
PSK Phase Shift Keying 相移键控
PSS Power System Stabilizer 电力系统稳定器
PT Potential Transformer 电压互感器
PTC Positive Temperature Coefficient 正温度系数
PWM Pulse Width Modulation 脉冲宽度调制

R

RC Resistance Capacitance 阻容
RCT Reverse Conducting Thyristor 逆导晶闸管
RRRV Rate of Rise of Recovery Voltage 恢复电压的上升速度
RTD Resistance Temperature Detector 电阻温度计

S

SBS Static Bypass Switch 自动静态旁路开关
SBD Schottky Barrier Diode 肖特基势垒二极管
SCC Single Chip Computer 单片微机
SCR Silicon Controlled Rectifier 可控硅整流器
SCR Short Circuit Ratio 短路比
SDI Serial Digital Interface 串行数字接口
SSC System Stabilizing Controller 系统稳定控制器
SHEPWM Selective Harmonic Elimination PWM 特定谐波消去 PWM
SVC Static Var Compensator 静止无功补偿器
SVG Static Var Generator 静止无功发生器
SVS Static Var System 静止无功补偿系统
SSR Solid State Relay 固态继电器
SLSI Super Large-Scale Integrated Circuit 超级大规模集成电路
SOA Safe Operating Area 安全工作区
SPDT Single Pole Double Throw 单刀双掷
SPST Single Pole Single Throw 单刀单掷
SPWM Sinusoidal PWM 正弦 PWM

SPIC Smart Power Integrated Circuit 智能功率集成电路
SIT Static Induction Transistor 静电感应晶体管
SITH Static Induction Thyristor 静电感应晶闸管

T

TCR Thyristor Controlled Reactor 晶闸管控制电抗器
TDR Time-Delay Relay 时间延时继电器
THD Total Harmonic Distortion 总谐波失真
TRV Transient Recovery Voltage 瞬态恢复电压
TRIAC Triode AC Switch 双向晶闸管
TTU Transformer Terminal Unit 变压器终端单元
TSC Thyristor Switched Capacitor 晶闸管投切电容器

U

UHF Ultra High Frequency 特高频
UHV Ultra High Voltage 超高压
UPS Uninterruptible Power Supply 不间断电源

V

VBR Variable Bit Rate 可变比特率
VCB Vacuum Circuit Breaker 真空断路器
VFC Voltage-to-Frequency Converter 电压-频率变换器
VLIS Very Large-Scale Integration 超大规模集成电路
VT Voltage Transformer 电压互感器
VMOS Vertical Metal-Oxide Semiconductor 虚拟金属氧化半导体
VVVF Variable Voltage Variable Frequency 变压变频
VSTL Voltage Source Type Inverter 电压（源）型逆变电路

Z

ZCS Zero Current Switching 零电流开关

参考文献

[1] BAYINDIR R, COLAK I, FULLI G, et al. Smart grid technologies and applications[M]. Amsterdam, Netherlands: Elsevier B.V, 2016.

[2] BOSE B K. Modern power electronics and AC drives [M]. Upper Saddle River, New Jersey: Prentice Holl, 2002.

[3] FLOYD T L. Digital fundamentals [M]. Upper Saddle River, New Jersey: Prentice Hall, 2012.

[4] FRANKLIN G F, POWELL J D, EMAMI-NAEINI A. Feedback control of dynamic systems (Sixth Edition) [M]. Pearson Education Malaysia, Pte. Ltd. Pearson Education, Inc., Upper Saddle River, New Jersey. 2009.

[5] HAYT W H, KEMMERLY J. Engineering circuit analysis[M]. New York: McGraw-Hill Science, 2006.

[6] MOHAN N, UNDELAND T M, ROBBINS W P. Power electronics[M]. Hoboken, N.J. J. Wiley, 2003.

[7] MOMOH J. Smart grid: fundamentals of design and analysis[M]. Hoboken, New Jersey: Wiley, 2012.

[8] NILSSON J W, RIEDEL S A. Introductory circuits for electrical and computer engineering [M]. Upper Saddle River, New Jersey: Prentice Hall, 2001.

[9] RASHID M H. Power electronics: circuits, devices and applications[M]. Upper Saddle River, New Jersey: Prentice Hall, 2003.

[10] STEPHEN J C. Electric machinery fundamentals[M]. New York: McGraw Hill Education, 2011.

[11] 姜书艳, 刘珊, 张昌华, 等 . 自动化专业英语 [M]. 北京：电子工业出版社, 2016.

[12] 刘慧娟, 范瑜 . Electric Machinery[M]. 北京：机械工业出版社, 2014.

[13] 刘芹 . 科技英语口译 [M]. 上海：上海外语教育出版社, 2014.

[14] 王爱君, 王爽, 胡章钰 . 科技英语阅读与翻译实用教程 [M]. 北京：新时代出版社, 2003.

[15] 王彩霞, 刘明华, 蔡宁, 等 . 自动化专业英语教程 [M]. 北京：机械工业出版社, 2015.

[16] 王宏文 . 自动化专业英语教程 [M]. 北京：机械工业出版社, 2015.

[17] 王军, 宋舒 . 自动化专业英语 [M]. 重庆：重庆大学出版社, 2016.

[18] 王兆安, 刘进军 . 电力电子技术 [M]. 北京：机械工业出版社, 2013.

[19] 张晓江, 孙慧芳, 黄云志, 等 . 自动化专业英语 [M]. 合肥：中国科学技术大学出版社, 2012.